NEXT GENERATION SEQUENCING & APPLICATIONS

NEXT GENERATION SEQUENCING & APPLICATIONS

Edited by:

Preethi Kartan

Delve PUBLISHING

www.delvepublishing.com

Next Generation Sequencing & Applications

Preethi Kartan

Delve Publishing
2010 Winston Park Drive,
2nd Floor
Oakville, ON L6H 5R7
Canada
www.delvepublishing.com
Tel: 001-289-291-7705
001-905-616-2116
Fax: 001-289-291-7601
Email: orders@arclereducation.com

ISBN: 978-1-77361-110-5 (Hardcover)

DECLARATION

Some content or chapters in this book are open access copyright free published research work, which is published under Creative Commons License and are indicated with the citation. We are thankful to the publishers and authors of the content and chapters as without them this book wouldn't have been possible.

ABOUT THE EDITOR

Preethi Kartan

Preethi is a postgraduate in Biotechnology from University of Leeds, UK. She is currently working as a Scientific Associate at one of the CRO's in India and her interest lie in life sciences related writing.

TABLE OF CONTENT

LIST OF CONTRIBUTORS

Preethi Kartan

Lin Liu, Yinhu Li
NGS Sequencing Department, Beijing Genomics Institute (BGI), 4th Floor, Building 11, Beishan Industrial Zone, Yantian District, Guangdong, Shenzhen 518083, China

Siliang Li
NGS Sequencing Department, Beijing Genomics Institute (BGI), 4th Floor, Building 11, Beishan Industrial Zone, Yantian District, Guangdong, Shenzhen 518083, China

Ni Hu
NGS Sequencing Department, Beijing Genomics Institute (BGI), 4th Floor, Building 11, Beishan Industrial Zone, Yantian District, Guangdong, Shenzhen 518083, China

Yimin He
NGS Sequencing Department, Beijing Genomics Institute (BGI), 4th Floor, Building 11, Beishan Industrial Zone, Yantian District, Guangdong, Shenzhen 518083, China

Ray Pong
NGS Sequencing Department, Beijing Genomics Institute (BGI), 4th Floor, Building 11, Beishan Industrial Zone, Yantian District, Guangdong, Shenzhen 518083, China

Danni Lin
NGS Sequencing Department, Beijing Genomics Institute (BGI), 4th Floor, Building 11, Beishan Industrial Zone, Yantian District, Guangdong, Shenzhen 518083, China

Lihua Lu
NGS Sequencing Department, Beijing Genomics Institute (BGI), 4th Floor, Building 11, Beishan Industrial Zone, Yantian District, Guangdong, Shenzhen 518083, China

Maggie Law
NGS Sequencing Department, Beijing Genomics Institute (BGI), 4th Floor, Building 11, Beishan Industrial Zone, Yantian District, Guangdong, Shenzhen 518083, China

Wilhelm J Ansorge
Ecole Polytechnique Federal Lausanne, EPFL, Lausanne, Switzerland

Tarek Hamed Attia
Pediatric Department, Faculty of Medicine, Zagazig University, Egypt

Maysaa Abdallah Saeed
Tropical Medicine Department, Faculty of Medicine, Zagazig University, Egypt

Chandra S Pareek
Functional Genomics Lab. Faculty of Biology and Environmental Protection, Nicolaus Copernicus University, Torun, Poland
Interdisciplinary Centre of Modern Technology, Nicolaus Copernicus University, Torun, Poland

Wenhui Laura Li
Department of Pathology and Laboratory Medicine, Children's Hospital Los Angeles, Los Angeles, California, USA

Suma J
New York Genome Center, New York, USA

Sivakumar G
Qassim University, College of Applied Medical Sciences, Buraidah, Kingdom of Saudi Arabia

Dhuha H Nawab
Pharmacy College, Umm Al Qura University, Saudi Arabia

Hubert E Blum
Department of Medicine, University Medical Center Freiburg, Germany

Konrad Oexle
Institute of Human Genetics, Technical University Munich, Germany

Edward J Fox
Departments of Pathology and Biochemistry, University of Washington, USA

Kate S Reid-Bayliss
Departments of Pathology and Biochemistry, University of Washington, USA

Mary J Emond
Department of Biostatistics, University of Washington, USA

Lawrence A Loeb
Departments of Pathology and Biochemistry, University of Washington, USA

Srilakshmi Srinivasan
Australian Prostate Cancer Centre- Queensland, Translational Research Institute, Queensland University of Technology, Woolloongabba, AustraliaSanger

Jyotsna Batra
Australian Prostate Cancer Centre- Queensland, Translational Research Institute, Queensland University of Technology, Woolloongabba, AustraliaSanger

Vathsala Palakkod Govindan
Indian Institute of Science, Bangalore, Karnataka, India

Krishna Murthy P
Indian Institute of Science, Bangalore, Karnataka, India

Jorge Munoz-Minjares
Department of Electronics Engineering, Guanajuato University, Salamanca, Mexico

Yuriy S Shmaliy
Department of Electronics Engineering, Guanajuato University, Salamanca, Mexico

Chee-Seng Ku
Department of Medical Epidemiology and Biostatistics, Karolinska Institutet, Stockholm, Sweden

Gillian M. Blue
Kids Heart Research, The Children's Hospital at Westmead, Sydney, Australia
Heart Centre for Children, The Children's Hospital at Westmead, Sydney, Australia
Sydney Medical School, University of Sydney, Australia

David S. Winlaw
Kids Heart Research, The Children's Hospital at Westmead, Sydney, Australia
Heart Centre for Children, The Children's Hospital at Westmead, Sydney, Australia
Sydney Medical School, University of Sydney, Australia

Thomas Mabey
Department of Biochemistry, Faculty of Medicine, Chulalongkorn University, King Chulalongkorn Memorial Hospital, Thailand

Sittisak Honsawek
Department of Biochemistry, Faculty of Medicine, Chulalongkorn University, King Chulalongkorn Memorial Hospital, Thailand

Fei Ye
Department of Cancer Biology, Vanderbilt University, USA

David C
Department of Cancer Biology, Vanderbilt University, USA

Samuels
Department of Cancer Biology, Vanderbilt University, USA

Yan Guo
Department of Cancer Biology, Vanderbilt University, USA

Kristen H. Taylor
University of Missouri-Columbia School of Medicine, Ellis Fischel Cancer Center, Columbia, MO 65212, USA

Huidong Shi
Medical College of Georgia, Augusta, GA 30912, USA

Charles W. Caldwell
University of Missouri-Columbia School of Medicine, Ellis Fischel Cancer Center, Columbia, MO 65212, USA

Thorfinn Sand Korneliussen
Centre for GeoGenetics, Natural History Museum of Denmark, Copenhagen, Denmark

Anders Albrechtsen
Bioinformatics Centre, Department of Biology, University of Copenhagen, Ole Maaloes Vej 5, DK-2200 Copenhagen, Denmark

Rasmus Nielsen
Centre for GeoGenetics, Natural History Museum of Denmark, Copenhagen, Denmark
Department of Integrative Biology and Statistics, UC-Berkeley, 4098 VLSB, Berkeley, California 94720, USA

Michael Knapp
Allan Wilson Centre for Molecular Ecology and Evolution, Department of Anatomy and Structural Biology, University of Otago, 270 Great King Street, 9016 Dunedin, New Zealand

Michael Hofreiter
Department of Biology, The University of York, Wentworth Way, Heslington, YO10 5DD, York, UK

PREFACE

Billions of spots of tiny genetic code comprise the human genome. It was DNA sequencing technology that had revolutionized genomic research by decoding the valuable genetic information by giving the picture of an exact order of occurrence of nucleotides in a DNA. The inception of first-generation sequencing method, also called Sanger sequencing took place in 1975. The first major breakthrough of first-generation sequencing comes, when the 13 year log Human Genome Project (HGP) was completed in 2003 at a cost $3 million. With ever increasing demands of researchers and clinicians, complex genomic research require a depth of information which is however beyond the capacity of traditional DNA sequencing technologies. These research questions gaps are very well addressed by Next-generation sequencing (NGS) has filled that gap of cheaper as well as faster sequencing technology.

It is just a decade old technology, but it has popularize the next-generation sequencing to high-throughput sequencing hat allow millions to trillions of observations to be made in parallel during a single instrument run. Since the introduction of these technologies, the number of applications and methods that influence the power of genome-scale sequencing has increased exponentially. Although in genome research NGS has mostly superseded conventional Sanger sequencing, it has not yet translated into routine clinical practice.

The following chapter will highlight the concepts, technologies, and methods of next-generation sequencing to illustrate the breadth and depth of the applications and research areas that are driving progress in genomics.

Editor

CHAPTER 1

INTRODUCTION

Preethi Kartan

INTRODUCTION TO NSG

Automated Sanger sequencing is considered the "*first-generation*" of DNA sequencing technologies. However; in 2000s, the concept of DNA sequencing underwent drastic changes. The Next Generation Sequencing (NGS) or the second generation sequencing methods are highly parallelized methods that enable to sequence thousands to millions of molecules simultaneously. With this technology, a huge amount of data (approx. hundreds of gigabases) is generated in a single sequential run which is analyzed through algorithm tools specific in analyzing the NGS data. The basic concept of NSG technology is analogous to conventional capillary electrophoreses (CE) based Sanger sequencing method--- During DNA synthesis, DNA polymerase catalyzes the incorporation of fluorescence labeled deoxyribonucleotides triphosphate (dNTPs) into a DNA template strand. Flourphore excitation identifies the incorporated nucleotide during each cycle of synthesis. NGS extend its process across massive parallel sequencing, where millions of fragments of DNA from a single sample can be accurately sequenced. The NGS, also known as

Massive Parallel Sequencing, allows the complete genome (of human) to be sequenced within a day (Reis-Filho, 2009).

The development under NGS gained momentum in late 20th and early 21st century. It was in between 2001 and 2006 when microarray technology offered the first genome-scale parallel analysis of DNA and RNA. In 2000, the first NGS technology, known as Massively Parallel Signature Sequencing (MPSS), was launched by a USA based Lynx Therapeutics Company. From there it had begun to emerge that permitted an unbiased means to examine billions of templates of DNA and RNA.

COMPARISON OF SANGER SEQUENCING WITH NEXT-GENERATION SEQUENCING

A typical Sanger sequencing involves fragmentation of genomic DNA and cloning to a plasmid vector in transfected bacteria. A transformed bacterial colony is picked for each sequencing reaction. Each sequencing reaction makes use of ddNTP-terminated, fluorescently labeled products which are subjected to high-resolution electrophoresis separation. As labeled fragments of discrete sizes pass a detector, a four-channel emission spectrum is used to generate a sequencing trace. In next-generation sequencing, genomic DNA is fragmented and then common adaptors are ligated to the fragments. The resulting sequencing library is then immobilized onto an array of millions of spatially separated PCR colonies, or 'Polonies'. As each nucleotide is added, image-based detection of fluorescent labels can be used to acquire sequencing data.

Why NSG?

The NSG technology is an indispensable and universal biological tool that touches every critical aspect of biological study i.e. relationship between genotype and the associated phenotypes. Some of the interesting properties which put it way ahead to conventional DNA sequencing are:

Cheaper and Faster

It provides us a much cheaper and faster alternative to traditional Sanger sequencing. The current electrophoretic methodology includes standard molecular biology techniques like genomic sample preparation and the separation of DNA fragments using capillary array electrophoresis, for sequencing an entire mammalian genome. It remains far too expensive ($5 million) to make genome sequencing ubiquitous (Hert et al., 2008). Researchers now have the chance to sequence whole small genome in a day with this non-electrophoresis-based sequencing tool. More than five human genomes can be sequenced in a single run, generating data in less than a week. However; it took over 10 years and cost nearly $3 billion to complete the Human Genome Project using capillary electrophoresis-based Sanger sequencing. The reduced time, manpower and reagents in NGS make it cost effective.

High-Throughput Sequencing

NGS is highly scalable and allow fine resolution to tune the level of resolution at specific targeted sequences of the genome. Target specific sequencing of specific genes or genomic region allows the researchers to discover genes and regulatory pathways associated with disease. High-throughput sequencing helps in better therapeutic decision-making for several genetic disorders, including many cancers (Grada&Weinbrecht, 2013).

Dynamic Range for Expression Profiling

NGS makes sequence-based gene expression analysis a "digital" alternative to analog techniques. It helps in the quantification of RNA expression with the breadth of a microarray and the resolution of qPCR. Instead of generating hundreds of longer sequence reads of 800–1000 bp, NGS technology produces shorter reads (40–400 bp) ranging on the order of gigabases (Gb) per run (Ansorje, 2009). RNA-Seq (NGS of RNA) provides the information of entire transcriptome information in the form of sequence, without any need of previous database related to genetic sequence of an organism.

Advantages of NSG over Microarray

Microarray technology has been widely used for the study of cancer genomes and transcriptome but some its promises have not materialized due to limitations of microarray data. Microarray design requires a previous set of database / knowledge of the genome or genomic characteristics. This directly influences the efficacy of array in cases of incomplete, incorrect, or outdated genome annotations. Furthermore, Microarrays poses restriction for metagenomic approaches, where the genetic content from undefined mixtures of organisms in an environment is sampled simultaneously in a group. However, for NSG prior knowledge of genome annotation is helpful, but not mandatory, rather data from next-generation platforms is assembled into a genome *de novo*

A major obstacle usually a problem in microarray analysis is cross-hybridization between similar sequences. This present a restriction for microarray analysis to the non-repetitive fraction of genomes and complicates analysis of related genes (or features), alternatively spliced transcripts, allelic gene variants, and SNPs. In fact in NSG, genetic material is directly sequenced and it removes experimental bias and cross-hybridization issues from the analysis. Since NGS offers precision at single-nucleotide, one can directly monitor the expression from gene that differs in as little as one nucleotide of sequence (Ari and Arikan, 2016).

Microarray gene expression measurement is limited by High signal to noise ratios. This makes the detection of low-abundance sequences difficult and quantitative resolution of changes in highly represented sequences equally challenging. In contrast, next-generation sequencing quantifies discrete signal from sequence-based approaches which is based on counting sequence tags rather than relative measures between samples as in microarray. It offers an unlimited quantitative dynamic range of signal, even for changes in rare and highly expressed sequences present in the same sample.

Considering the quantity of sample required, micrograms of DNA are needed to hybridize to arrays, esp. for whole genome tiling arrays represented on multiple slides. However, nanograms of genetic material are sufficient for NGS, thus eliminating the dependence on PCR amplification can introduce bias into samples.

The variety of available microarray formats, preparative methodologies and analytical approaches may limit the reproducibility of microarray data by only giving a semi-quantitative assessment of gene expression on a genome-wide scale. Data output in the form of tally of RNA sequences in next-generation platforms discover novel RNA variants and splice sites and provides a strong alternative approach for gene expression studies (Raza and Ahmad, 2016).

METHODOLOGY

In massively parallel sequencing or NSG, the process is a stepwise reaction series that consists of (a) a nucleotide addition step, (b) a detection step that determines the identity of the incorporated nucleotides on each fragment focus being sequenced, and (c) a wash step that may include chemistry to remove fluorescent labels or blocking groups. In essence, next-generation sequencing instruments conduct sequencing and detection simultaneously rather than as distinct processes; detection of previous sample is completed before the other sequencing takes place. All these steps are performed in a format or array that allows hundreds of thousands to billions of reaction foci to be sequenced with each instrument run and, generating enormous data sets.

The next-generation sequencing instruments do not require a cloning step in contrast to first generation sequencing. Rather, the DNA to be sequenced is used to construct a library of fragments that have synthetic DNAs (adapters) added covalently to each fragment end by use of DNA ligase. These adapters are universal sequences, specific/ unique to each platform, which can be used to polymerase-amplify the library fragments during further steps of the process.

Sequence Library Preparation

The starting material for all NGS experiments is double-stranded DNA; however, the source may vary (e.g., genomic DNA, reverse-transcribed RNA or cDNA, immunoprecipitated DNA). All the existing sequencing platforms require some level of pre-processing of DNA into a DNA library suitable for sequencing. The sequential steps are 1) shearing of high molecular weight DNA into an appropriate Platform-specific size range, 2) end polishing step to generate blunt ended DNA fragments 3) Specific adapters are ligated to these fragments by either A/T overhang or direct blunt ligation 3′ and 5′ ends to make the library functional, 4) pre-amplification step prior to sequencing. Life Technologies (Solid, PGM, Proton) libraries require a nick translation step to get functional molecules following adapter ligation; while other technologies (Illumnia), the DNA sample is in ready stage for loading (Buermans and Dunnen, 2014).

Commercially Available NSG/ Next Generation Sequencing Technologies

The different sequence platform vendors have devised different strategies to prepare the sequence libraries into suitable templates as well as to detect the signal and ultimately read the DNA sequence. The method used for determining template DNA sequence can be used to classify NGS techniques as follows:

Illumina Sequencing

It is one of the major platforms introduced in 2006 for High through put sequencing. All of the enzymatic processes and imaging steps of the Illumina technology takes place in a flow cell with a field of oligos attached, instead of a chip containing individual microwells with beads. Comparatively longer fragments are considered from the template library for ligation using generic adaptors. Further, these ligated fragments are annealed (attached) to a glass slide. In the next stage, amplification is carried with PCR. Later, amplified copies of reads are separated into single strands, which is to be sequenced. In this method the dye-termination approach is used which is different from Sanger technology, however, in that the dye terminators are reversible, so they are removed after each imaging cycle to allows incorporation of one base per cycle into each template molecule. Next, the slide is flooded with fluorescently labelled Reversible dye-terminator nucleotides dNTPS with unique color for identification, over the flow cell (Hodkinson, and Grice, 2015). The fluorescent signal in form of image is recorded, which indicates the addition of the base per extension step for each cluster. Terminators are reversed, so that the process can repeat and nucleotides can continue to be added in subsequent cycles. At the same time, fluorescent signal is also removed in order to prevent future contaminations of the signals. Using the computers, the bases are detected at each site in each image, which in turn help in construction of a sequence.

Ion Torrent PGM sequencing

A completely different approach to next-generation sequencing is embodied in an instrument system, commercialized in 2010 by Ion Torrent, which detects the release of H^+ ions, a by-product of nucleotide incorporation which is quantified as a change in pH through a novel coupled silicon detector (Mardis, 2013).

For this approach, library construction includes DNA fragmentation (template DNA or RNA is fragmented into size of ~200bp.), enzymatic end polishing, and adapter ligation. But amplification of library fragments occurs by a unique approach known as emulsion PCR, which quantities the library fragments and dilutes them to be mixed in equimolar quantities with small beads, PCR reactants, and DNA polymerase molecules. The beads have covalently linked adapter complementary sequences on their surfaces to facilitate amplification on the bead/ spheres. Enriched beads are primed for sequencing by annealing a sequencing primer and are deposited into the wells of an Ion Chip, a specialized silicon chip designed to detect pH changes (addition of dNTP) within individual wells of the sequencer, as the reaction progresses stepwise.

454 Pyrosequencing

In 2005, Life Sciences (a Roche company) brought the first next-generation sequencing technologies to market with its special feature of long reads especially for studies of microbiomes, since longer reads can generally be identified with greater accuracy and precision. The overall approach for 454 is based on pyrosequencing. Instead of using dideoxynucleotides to terminate the chain amplification, pyrosequencing technology relies on the detection of pyrophosphate released during nucleotide (dNTPS) incorporation. The library DNAs with 454-specific adaptors are denatured into single strand and captured by amplification beads followed by emulsion PCR. On a fiber-optic chip filled with a field of microwells, known as a PicoTiterPlate, , one of dNTP (dATP, dGTP, dCTP, dTTP) will complement to the bases of the template strand with the help of ATP sulfurylase, luciferase, luciferin, DNA polymerase, and adenosine 5′ phosphosulfate (APS) and release pyrophosphate (PPi) which equals the amount of incorporated nucleotide. The ATP transformed from PPi drives the luciferin into oxyluciferin and generates visible light which is recorded to infer the sequence of the DNA fragments in each well Hodkinson, and Grice,2015).. 454 sequencing technology has the ability to sequence much longer reads as compared to Illumina.

Applied Biosystems SOLiD

SOLiD (Sequencing by Oligo Ligation Detection) was introduced by Applied Biosystems in 2006. Its utility has been demonstrated for microbiome applications. The sequencer adopts the technology of two-base sequencing based on ligation sequencing. DNA ligase is used for incorporation, making it a "sequencing-by-ligation" approach, as opposed to the "sequencing-by-synthesis" approaches mentioned previously (Mardis, 2008). The SOLiD 5500xl W Genetic Analyzer produces up to 3 billion reads per run with reads that are 75 bases long. Although it can generate large numbers of sequences in a run, the persistent short read length has greatly limited its utility (Mitra et al., 2013).

Single-Molecule Next-Generation Sequencing

Although Helicos BioSciences produced the first single-molecule sequencing platform, Pacific Biosciences, Inc., has had much greater commercial success and currently leads the way for single-molecule sequencing. Pacific Biosciences (PacBio) constitutes a single-molecule, real-time sequencing approach which combines nanotechnology with molecular biology and highly sensitive fluorescence detection to achieve single-molecule DNA sequencing. The PacBio sequencing system involves no amplification step, and the prepared library molecule is the sequencing template; setting it apart from the other major next-generation sequencing systems. So it simplifies the DNA sample preparation process and avoids PCR induced errors (Buermans and Dunnen, 2014).

The sequencing is performed on a chip containing many zero-mode waveguide (ZMW) detectors. DNA polymerases are attached to the ZMW detectors and phospholinked fluorescence labeled nucleotide incorporation is imaged in real time as new DNA strands are synthesized. Once fluorescent nucleotides are added to the chip surface, the instrument optics continually review the active site of each ZMW-bound polymerase in real time, and detect the signal from its laser light– stimulated emission (Mardis, 2013).

A comparative view of five major next-generation sequencing platform is shown in Table 1

Table 1: Outline of the five major next-generation sequencing platform families

Sequencing Platform	Clonal Amplification	Sequencing Chemistry	Highest Average Read Length
Roch 454	Emulsion PCR	Pyrosequencing (seq-by-synthesis)	700 bp (paired-end sequencing available)
Illumina	Bridge amplification	Reversible dye terminator (seq-by-synthesis)	300 bp (overlapping paired-end sequencing available)
ABI SOLiD System	Emulsion PCR	Oligonucleotide 8-mer chained ligation (seq-by-ligation)	75 bp (paired-end sequencing available)
Ion Torrent	Emulsion PCR	Proton(H+) detection (seq-by-synthesis)	400 bp (bidirectional sequencing available)
PacBio	N/A (single molecule)	Phospholinked fluorescent nucleotides (Single molecule realtime (SMRT) sequencing)	8,500 bp

APPLICATION AREAS OF NGS TECHNOLOGY

The next-generation sequencing technologies offer novel and quick ways for genome-wide characterization and profiling of mRNAs, small RNAs, transcription factor regions, structure of chromatin and DNA methylation patterns, and metagenomics. (Ansorge , 2009). The next generation sequencing is now exploring several aspects of medical and biochemical researches worldwide. Below we discuss some widely used applications of Next generation sequencing.

Whole Genome Sequencing – De Novo Sequencing

The main application of NGS technology is the complete characterization of the entire genome of a particular species. The use of NGS in the *de novo* sequencing of eukaryotes is currently expanding, with uptake in cDNA sequencing studies. Next Generation Sequencing performs massively parallel sequencing which facilitates the high-throughput sequencing and thereby responsible for entire genome sequencing in less than a day in minimum cost. Human genome sequencing helps in identifying the specific genes and their regulatory partners involved in the pathological processes (Grada and Weinbrecht, 2013).The major applications of whole genome sequencing constitutes areas of Biodiversity, epidemiology, pathogen detection, evolution.

Whole Exome Sequencing

It helps in investigating the protein coding regions within the genome. Exome shares only small portion of the whole genome and it remains a critical site of concern for clinicians and researchers. Mutations, alterations in this part of genome results are possible causes of a disease or other abnormal conditions. In case of human genome, exome represents less than 2% of the total genome, but contains nearly 85% of disease-causing variants (Rizzo and Buck, 2012). Therefore, sequencing only exome part of genome, in spite of whole genome sequencing results in the cost-effective sequencing. Important applications of exome sequencing involve variant association studies, discovering relevant mutations in exome region and their expressed phenotype, identification of disease-related variations and diagnosis, and cancer clonality.

Targeted Resequencing

Until recently, re-sequencing has been limited to studies on small numbers of genetic loci. NGS is currently being used to re-sequence bacterial isolates to provide a better understanding of their genomic diversity and the evolution of pathogens, and also to identify mutant alleles and structural variants in human, plant and animal genomes. It involves in the sequencing of genomic regions of interest which can be a small subset of the genome, such as exome, a set of genes, a particular chromosome or a region of particular interest; which is a cost effective approach. Moreover, narrowing down the explored region enables one to get closer variation analysis which could have been otherwise impossible, when whole genome analysis was the only alternative. Comparing the genomes of multiple individuals/ strains in a species will enable detailed characterization which link quantitative variation in phenotype to genotype, and allowing an understanding of the molecular basis for adaptation. Resequencing has been found useful in discovering the point mutations, insertion or deletions, gene rearrangements and variations occurring in copy number.

Metagenomics

Metagenomic approaches identify genomic signatures from mixed populations of microorganisms, many of which may be unculturable. These signatures can act as markers to detect the cause of any infectious disease. Analyzing all DNA content from control versus infected (diseased) individuals will reveal the "extra" DNA base responsible for pathogenicity. Metagenomics can be performed by a sequence it all approach or by focusing on specific uniformly conserved sequences like e.g. ribosomal RNA genes only. Metagenomic approach enables the study of organisms that is not culturable esp for microbial communities in human gut environment. Based on DNA sequencing one gets an idea of the complexity and constitution of entire ecosystems (Ercolini, 2012)

RNA Sequencing and Expression Profiling Analysis

High-throughput sequencing has been applied to study RNA transcripts and is referred to as RNA-seq or transcriptome-seq, and has provided broad information on gene expression levels and differentially expressed transcripts among groups of samples. After the identical preprocessing procedures as in DNA-seq data analysis, RNA-seq data can be used for *de novo* transcriptome assembly, expression profiling analysis, and transcriptome epigenetics (Lee et al., 2013).

While *de novo* **DNA assembly is aimed toward building genomic platform for novel species without any prior reference,** *de novo* **assembly of RNA-**seq data sketches an overview and extracts clues to the "transcriptome. Many classes of small RNAs (sRNAs), such as miRNA, piwi-interacting RNA (piRNA), and small interfering RNA (siRNA), are important in post-translational regulation of gene expression.. Discovery of new sRNAs is highly facilitated by NGS technology via its massively parallel high throughput of sequencing, which makes it possible to detect sRNAs with lower expression that are hard to find by traditional Sanger sequencing with high sensitivity (Maclean et al., 2009 and Lee et al., 2013). NGS also provides a new way to explore the role of microRNAs in virus replication and pathogenesis.

Chromatin Immunoprecipitation Sequencing (ChIP-Seq)

It is another high-throughput method to analyze the interactions of proteins with DNA and RNA involved in gene regulation. ChIP-Seq combines the ChIP technique and NGS technique to identify the binding sites of the transcription factors and associated proteins on a whole genome scale (Jhonson et al., 2007). The application of next-generation sequencing (NGS) to ChIP enables in interpreting the regulation events which may play a role in various diseases and biological pathways, such as gene regulation, DNA repair, DNA synthesis and abnormal expression levels. Important applications of ChIP-Seq include analysis of epigenetic patterns without requiring the probe; providing unbiased genome-wide profiling.

CHALLENGES OF NSG.

With all the advantages and potential NGS brings to research and diagnostics, it also has several drawbacks which need to be addressed. The major issue is the too much dependence on bioinformatics and computational science to generate alignment protocols. There is also a real concern regarding the amount of data storage capacity required (per sequencing run on NGS platforms runs into the gigabytes) and then analyze the massive amounts of data generated from every sequencing run (Behjati and Tarpey, 2013). In addition, data safety poses a significant obstacle, as genetic data is key to health information. Generation of large amounts of sequence data, thus, questions the right of privacy and confidentiality for an individual.

Other challenges faced are that of the costs associated with gaining the equipment, software and consumables required for NGS. Current estimates of the start-up capital required are around \$500,000 (~£300,000) but this can be offset by the cost per base of sequencing as compared to Sanger method which is more expensive and a much larger data output is also achieved.

CONCLUSION

It is said that the application of next-generation sequencing will be sufficiently fast, accurate and cheap in near future to be used in routine clinical microbiology practice, where it could replace many complex current techniques with a single, and more efficient workflow. On the other hand, plenty of technical challenges, which need to be deeply analyzed and solved, emerged with these technologies. Every sequencing generation and platform, by reason of its methodological approach, carries characteristic advantages and disadvantages which determine the fitness for certain applications. Thus, assessment of these features, limitations, and potential applications help shaping the studies that will determine the route of omic technologies.

REFERENCES

1. Ari, Ş., & Arikan, M. (2016). Next-Generation Sequencing: Advantages, Disadvantages, and Future. In Plant Omics: Trends and Applications (pp. 109-135).
2. Ansorge, W.J. (2009). Next-generation DNA sequencing techniques, 25(4): 195-203.
3. Buermans HP, den Dunnen JT. 2014. Next generation sequencing technology: Advances and applications. Bochim Biophys Acta. 1842(10):1932-1941.
4. Behjati, S., & Tarpey, P. S. (2013). What is next generation sequencing? *Archives of Disease in Childhood. Education and Practice Edition*, *98*(6), 236–238.
5. D. Ercolini, F. De Filippis, A. La Storia, M. Iacono. 2012. Remake by high-throughput sequencing of the microbiota involved in the production of water buffalo mozzarella cheese, Appl. Environ. Microbiol. 78 : 8142–8145.
6. Hert DG, Fredlake CP, Barron AE.2008. Advantages and limitations of next-generation sequencing technologies: a comparison of electrophoresis and non-electrophoresis methods. Electrophoresis, **29**(23):4618-26.
7. Grada, A., & Weinbrecht, K. (2013). Next-generation sequencing: methodology and application. Journal of Investigative Dermatology,

133(8), e11

8. Hodkinson, B. P., & Grice, E. A. (2015). Next-Generation Sequencing: A Review of Technologies and Tools for Wound Microbiome Research. *Advances in Wound Care*, *4*(1), 50–58.
9. Johnson, D. S., Mortazavi, A., Myers, R. M. & Wold B. (2007). Genome-wide mapping of in vivo protein–DNA interactions. Science 316, 1497–1502
10. Kim, R. Y., Xu, H., Myllykangas, S., & Ji, H. (2011). Genetic-based biomarkers and next-generation sequencing: the future of personalized care in colorectal cancer. *Personalized Medicine*, *8*(3), 331–345.
11. Lee CH, Chiu YC, Wang LB, Kuo YL, Chuang EY, Lai LC, Tsai MH.2013. Common applications of nextgeneration sequencing technologies in genomic research. Transl Cancer Res 2(1):33-45
12. MacLean D[1], Jones JD, Studholme DJ. 2009. Application of 'next-generation' sequencing technologies to microbial genetics. Nat Rev Microbiol, 7(4):287-96.
13. Mardis ER. 2013. Next-Generation Sequencing Platforms Annu. Rev. Anal. Chem.. 6:287–303
14. Mardis,ER . 2008. "The impact of next-generation sequencing technology on genetics," Trends in Genetics, vol. 24, no. 3, pp. 133–141,
15. Michael L. Metzker. Sequencing technologies — the next generation *Nature Reviews Genetics* 11, 31-46 (January 2010)
16. Mitra, S., Förster-Fromme, K., Damms-Machado, A., Scheurenbrand, T., Biskup, S., Huson, D. H., & Bischoff, S. C. (2013). Analysis of the intestinal microbiota using SOLiD 16S rRNA gene sequencing and SOLiD shotgun sequencing. *BMC Genomics*, *14*(Suppl 5), S16.
17. Paul J. Hurd, Christopher J. Nelson; Advantages of next-generation sequencing versus the microarray in epigenetic research. *Brief Funct Genomics* 2009; 8 (3): 174-183.
18. Raza& Ahmad 2016. Principle, analysis, application and challenges of next-generation sequencing: a review
19. Reis-Filho, J. S. (2009). Next-generation sequencing. *Breast Cancer Research : BCR*, *11*(Suppl 3), S12.

20. Rizzo, JM and Buck MJ. 2012. Key Principles and Clinical Applications of *"Next-Generation"* DNA Sequencing Cancer Prev Res; 5(7); 887–900.

CHAPTER 2

COMPARISON OF NEXT-GENERATION SEQUENCING SYSTEMS

Lin Liu, Yinhu Li, Siliang Li, Ni Hu, Yimin He, Ray Pong, Danni Lin, Lihua Lu, and Maggie Law

NGS Sequencing Department, Beijing Genomics Institute (BGI), 4th Floor, Building 11, Beishan Industrial Zone, Yantian District, Guangdong, Shenzhen 518083, China

ABSTRACT

With fast development and wide applications of next-generation sequencing (NGS) technologies, genomic sequence information is within reach to aid the achievement of goals to decode life mysteries, make better crops, detect pathogens, and improve life qualities. NGS systems are typically represented by SOLiD/Ion Torrent PGM from Life Sciences, Genome Analyzer/HiSeq 2000/MiSeq from Illumina, and GS FLX Titanium/GS Junior from Roche. Beijing Genomics Institute (BGI), which possesses the world's biggest sequencing capacity, has multiple NGS systems including 137 HiSeq 2000, 27 SOLiD, one Ion Torrent PGM, one MiSeq, and one 454 sequencer. We have accumulated extensive experience in sample handling, sequencing, and bioinformatics analysis. In this paper, technologies of these systems are reviewed, and first-hand data from extensive experience is summarized and analyzed to discuss the advantages and specifics associated with each sequencing system. At last, applications of NGS are summarized.

INTRODUCTION

(Deoxyribonucleic acid) DNA was demonstrated as the genetic material by Oswald Theodore Avery in 1944. Its double helical strand structure composed of four bases was determined by James D. Watson and Francis Crick in 1953, leading to the central dogma of molecular biology. In most cases, genomic DNA defined the species and individuals, which makes the DNA sequence fundamental to the research on the structures and functions of cells and the decoding of life mysteries [1]. DNA sequencing technologies could help biologists and health care providers in a broad range of applications such as molecular cloning, breeding, finding pathogenic genes, and comparative and evolution studies. DNA sequencing technologies ideally should be fast, accurate, easy-to-operate, and cheap. In the past thirty years, DNA sequencing technologies and applications have undergone tremendous development and act as the engine of the genome era which is characterized by vast amount of genome data and subsequently broad range of research areas and multiple applications. It is necessary to look back on the history of sequencing technology development to review the NGS systems (454, GA/HiSeq, and SOLiD), to compare their advantages and disadvantages, to discuss the various applications, and to evaluate the recently introduced PGM (personal genome machines) and third-generation sequencing technologies and applications. All of these aspects will be described in this paper. Most data and conclusions are from independent users who have extensive first-hand experience in these typical NGS systems in BGI (Beijing Genomics Institute).

Before talking about the NGS systems, we would like to review the history of DNA sequencing briefly. In 1977, Frederick Sanger developed DNA sequencing technology which was based on chain-termination method (also known as Sanger sequencing), and Walter Gilbert developed another sequencing technology based on chemical modification of DNA and subsequent cleavage at specific bases. Because of its high efficiency and low radioactivity, Sanger sequencing was adopted as the primary technology in the "first generation" of laboratory and commercial sequencing applications [2]. At that time, DNA sequencing was laborious and radioactive materials were required. After years of improvement, Applied Biosystems introduced the first automatic sequencing machine (namely AB370) in 1987, adopting capillary electrophoresis which made the sequencing faster and more accurate. AB370 could detect 96 bases one time, 500 K bases a day, and the read length could reach 600 bases. The current model AB3730xl can output 2.88 M bases per day and read length could reach 900 bases since 1995. Emerged in 1998, the automatic sequencing instruments and associated software using the capillary sequencing machines and Sanger sequencing technology became the main tools for the completion of human genome project in 2001 [3]. This project greatly stimulated the development of powerful novel sequencing instrument to increase speed and accuracy, while simultaneously reducing cost and manpower. Not only this, X-prize also accelerated the development of next-generation sequencing (NGS) [4]. The NGS technologies are different from the Sanger method in aspects of massively parallel analysis, high throughput, and reduced cost. Although NGS makes genome sequences handy, the followed data analysis and biological explanations are still the bottle-neck in understanding genomes.

Following the human genome project, 454 was launched by 454 in 2005, and Solexa released Genome Analyzer the next year, followed by (Sequencing by Oligo Ligation Detection) SOLiD provided from Agencourt, which are three most typical massively parallel sequencing systems in the next-generation sequencing (NGS) that shared good performance on throughput, accuracy, and cost compared with Sanger sequencing (shown in Table 1(a)). These founder companies were then purchased by other companies: in 2006 Agencourt was purchased by Applied Biosystems, and in 2007, 454 was purchased by Roche, while Solexa was purchased by Illumina. After years of evolution, these three systems exhibit better performance and their own advantages in terms of read length, accuracy, applications, consumables, man power requirement and informatics infrastructure, and so forth. The comparison of these three systems will be focused and discussed in the later part of this paper (also see Tables 1(a), 1(b), and 1(c)).

Table 1: (a) Advantage and mechanism of sequencers. (b) Components and cost of sequencers. (c) Application of sequencers

(a)

Sequencer	454 GS FLX	HiSeq 2000	**SOLiDv4**	**Sanger 3730xl**
Sequencing mechanism	Pyrosequencing	Sequencing by synthesis	Ligation and two-base coding	Dideoxy chain termination
Read length	700 bp	50SE, 50PE, 101PE	50 + 35 bp or 50 + 50 bp	400 ~ 900 bp
Accuracy	99.9%*	98%, (100PE)	99.94% *raw data	99.999%
Reads	1 M	3 G	1200~1400 M	—
Output data/run	0.7 Gb	600 Gb	120 Gb	1.9~84 Kb
Time/run	24 Hours	3~10 Days	7 Days for SE 14 Days for PE	20 Mins~3 Hours
Advantage	Read length, fast	High throughput	Accuracy	High quality, long read length
Disadvantage	Error rate with polybase more than 6, high cost, low throughput	Short read assembly	Short read assembly	High cost low throughput

(b)

Sequencers	**454 GS FLX**	**HiSeq 2000**	**SOLiDv4**	**3730xl**
Instrument price	Instrument $500,000, $7000 per run	Instrument $690,000, $6000/(30x) human genome	Instrument $495,000, $15,000/100 Gb	Instrument $95,000, about $4 per 800 bp reaction
CPU	2* Intel Xeon X5675	2* Intel Xeon X5560	8* processor 2.0 GHz	Pentium IV 3.0 GHz
Memory	48 GB	48 GB	16 GB	1 GB
Hard disk	1.1 TB	3 TB	10 TB	280 GB
Automation in library preparation	Yes	Yes	Yes	No
Other required device	REM e system	cBot system	EZ beads system	No
Cost/million bases	$10	$0.07	$0.13	$2400

(c)

Sequencers	454 GS FLX	HiSeq 2000	SOLiDv4	3730xl
Resequencing		Yes	Yes	
De novo	Yes	Yes		Yes
Cancer	Yes	Yes	Yes	
Array	Yes	Yes	Yes	Yes
High GC sample	Yes	Yes	Yes	
Bacterial	Yes	Yes	Yes	
Large genome	Yes	Yes		
Mutation detection	Yes	Yes	Yes	Yes

(1) All the data is taken from daily average performance runs in BGI. The average daily sequence data output is about 8 Tb in BGI when about 80% sequencers (mainly HiSeq 2000) are running.

(2) The reagent cost of 454 GS FLX Titanium is calculated based on the sequencing of 400 bp; the reagent cost of HiSeq 2000 is calculated based on the sequencing of 200 bp; the reagent cost of SOLiDv4 is calculated based on the sequencing of 85 bp.

(3) HiSeq 2000 is more flexible in sequencing types like 50SE, 50PE, or 101PE.

(4) SOLiD has high accuracy especially when coverage is more than 30x, so it is widely used in detecting variations in resequencing, targeted resequencing, and transcriptome sequencing. Lanes can be independently run to reduce cost.

ROCHE 454 SYSTEM

Roche 454 was the first commercially successful next generation system. This sequencer uses pyrosequencing technology [5]. Instead of using dideoxynucleotides to terminate the chain amplification, pyrosequencing technology relies on the detection of pyrophosphate released during nucleotide incorporation. The library DNAs with 454-specific adaptors are denatured into single strand and captured by amplification beads followed by emulsion PCR [6]. Then on a picotiter plate, one of dNTP (dATP, dGTP, dCTP, dTTP) will complement to the bases of the template strand with the help of ATP sulfurylase, luciferase, luciferin, DNA polymerase, and adenosine 5′ phosphosulfate (APS) and release pyrophosphate (PPi) which equals the amount of incorporated nucleotide. The ATP transformed from PPi drives the luciferin into oxyluciferin and generates visible light [7]. At the same time, the unmatched bases are degraded by apyrase [8]. Then another dNTP is added into the reaction system and the pyrosequencing reaction is repeated.

The read length of Roche 454 was initially 100–150 bp in 2005, 200000+ reads, and could output 20 Mb per run [9, 10]. In 2008 454 GS FLX Titanium system was launched; through upgrading, its read length could reach 700 bp with accuracy 99.9% after filter and output 0.7 G data per run within 24 hours. In late 2009 Roche combined the GS Junior a bench top system into the 454 sequencing system which simplified the library preparation and data processing, and output was also upgraded to 14 G per run [11, 12]. The most outstanding advantage of Roche is its speed: it takes only 10 hours from sequencing start till completion. The read length is also a distinguished character compared with other NGS systems (described in the later part of this paper). But the high cost of reagents remains a challenge for Roche 454. It is about $1 2 . 5 6 × 1 0− 6 per base (counting reagent use only). One of the shortcomings is that it has relatively high error rate in terms of poly-bases longer than 6 bp. But its library construction can be automated, and the emulsion PCR can be semiautomated which could reduce the manpower in a great extent. Other informatics infrastructure and sequencing advantages are listed and compared with HiSeq 2000 and SOLiD systems in Tables 1(a), 1(b), and 1(c).

454 GS FLX Titanium Software

GS RunProcessor is the main part of the GS FLX Titanium system. The software is in charge of picture background normalization, signal location correction, cross-talk correction, signals conversion, and sequencing data generation. GS RunProcessor would produce a series of files including SFF (standard flowgram format) files each time after run. SFF files contain the basecalled sequences and corresponding quality scores for all individual, high-quality reads (filtered reads). And it could be viewed directly from the screen of GS FLX Titanium system. Using GS De Novo Assembler, GS Reference Mapper and GS Amplicon Variant Analyzer provided by GS FLX Titanium system, SFF files can be applied in multiaspects and converted into fastq format for further data analyzing.

AB SOLID SYSTEM

(Sequencing by Oligo Ligation Detection) SOLiD was purchased by Applied Biosystems in 2006. The sequencer adopts the technology of two-base sequencing based on ligation sequencing. On a SOLiD flowcell, the libraries can be sequenced by 8 base-probe ligation which contains ligation site (the first base), cleavage site (the fifth base), and 4 different fluorescent dyes (linked to the last base) [10]. The fluorescent signal will be recorded during the probes complementary to the template strand and vanished by the cleavage of probes' last 3 bases. And the sequence of the fragment can be deduced after 5 round of sequencing using ladder primer sets.

The read length of SOLiD was initially 35 bp reads and the output was 3 G data per run. Owing to two-base sequencing method, SOLiD could reach a high accuracy of 99.85% after filtering. At the end of 2007, ABI released the first SOLiD system. In late 2010, the SOLiD 5500xl sequencing system was released. From SOLiD to SOLiD 5500xl, five upgrades were released by ABI in just three years. The SOLiD 5500xl realized improved read length, accuracy, and data output of 85 bp, 99.99%, and 30 G per run, respectively. A complete run could be finished within 7 days. The sequencing cost is about $\$ 40 \times 10^{-9}$ per base estimated from reagent use only by BGI users. But the short read length and resequencing only in applications is still its major shortcoming [13]. Application of SOLiD includes whole genome resequencing, targeted resequencing, transcriptome research (including gene expression profiling, small RNA analysis, and whole transcriptome analysis), and epigenome (like ChIP-Seq and methylation). Like other NGS systems, SOLiD's computational infrastructure is expensive and not trivial to use; it requires an air-conditioned data center, computing cluster, skilled personnel in computing, distributed memory cluster, fast networks, and batch queue system. Operating system used by most researchers is GNU/LINUX. Each solid sequencer run takes 7 days and generates around 4 TB of raw data. More data will be generated after bioinformatics analysis. This information is listed and compared with other NGS systems in Tables 1(a), 1(b), and 1(c). Automation can be used in library preparations, for example, Tecan system which integrated a Covaris A and Roche 454 REM e system [14].

SOLiD Software

After the sequencing with SOLiD, the original sequence of color coding will be accumulated. According to double-base coding matrix, the original color sequence can be decoded to get the base sequence if we knew the base types for one of any position in the sequence. Because of a kind of color corresponding four base pair, the color coding of the base will directly influence the decoding of its following base. It said that a wrong color coding will cause a chain decoding mistakes. BioScope is SOLiD data analysis package which provides a validated, single framework for resequencing, ChIP-Seq, and whole transcriptome analysis. It depends on reference for the follow-up data analysis. First, the software converts the base sequences of references into color coding sequence. Second, the color-coding sequence of references is compared with the original sequence of color-coding to get the information of mapping with newly developed mapping algorithm MaxMapper.

ILLUMINA GA/HISEQ SYSTEM

In 2006, Solexa released the Genome Analyzer (GA), and in 2007 the company was purchased by Illumina. The sequencer adopts the technology of sequencing by synthesis (SBS). The library with fixed adaptors is denatured to single strands and grafted to the flowcell, followed by bridge amplification to form clusters which contains clonal DNA fragments. Before sequencing, the library splices into single strands with the help of linearization enzyme [10], and then four kinds of nucleotides (ddATP, ddGTP, ddCTP, ddTTP) which contain different cleavable fluorescent dye and a removable blocking group would complement the template one base at a time, and the signal could be captured by a (charge-coupled device) CCD.

At first, solexa GA output was 1 G/run. Through improvements in polymerase, buffer, flowcell, and software, in 2009 the output of GA increased to 20 G/run in August (75PE), 30 G/run in October (100PE), and 50 G/run in December (Truseq V3, 150PE), and the latest GAIIx series can attain 85 G/run. In early 2010, Illumina launched HiSeq 2000, which adopts the same sequencing strategy with GA, and BGI was among the first globally to adopt the HiSeq system. Its output was 200 G per run initially, improved to 600 G per run currently which could be finished in 8 days. In the foreseeable future, it could reach 1 T/run when a personal genome cost could drop below $1 K. The error rate of 100PE could be below 2% in average after filtering (BGI's data). Compared with 454 and SOLiD, HiSeq 2000 is the cheapest in sequencing with $0.02/million bases (reagent counted only by BGI). With multiplexing incorporated in P5/P7 primers and adapters, it could handle thousands of samples simultaneously. HiSeq 2000 needs (HiSeq control software) HCS for program control, (real-time analyzer software) RTA to do on-instrument base-calling, and CASAVA for secondary analysis. There is a 3 TB hard disk in HiSeq 2000. With the aid of Truseq v3 reagents and associated softwares, HiSeq 2000 has improved much on high GC sequencing. MiSeq, a bench top sequencer launched in 2011 which shared most technologies with HiSeq, is especially convenient for amplicon and bacterial sample sequencing. It could sequence 150PE and generate 1.5 G/run in about 10 hrs including sample and library preparation time. Library preparation and their concentration measurement can both be automated with compatible systems like Agilent Bravo, Hamilton Banadu, Tecan, and Apricot Designs.

HiSeq Software

HiSeq control system (HCS) and real-time analyzer (RTA) are adopted by HiSeq 2000. These two softwares could calculate the number and position of clusters based on their first 20 bases, so the first 20 bases of each sequencing would decide each sequencing's output and quality. HiSeq 2000 uses two lasers and four filters to detect four types of nucleotide (A, T, G, and C). The emission spectra of these four kinds of nucleotides have cross-talk, so the images of four nucleotides are not independent and the distribution of bases would affect the quality of sequencing. The standard sequencing output files of the HiSeq 2000 consist of *bcl files, which contain the base calls and quality scores in each cycle. And then it is converted into *_qseq.txt files by BCL Converter. The ELAND program of CASAVA (offline software provided by Illumina) is used to match a large number of reads against a genome.

In conclusion, of the three NGS systems described before, the Illumina HiSeq 2000 features the biggest output and lowest reagent cost, the SOLiD system has the highest accuracy [11], and the Roche 454 system has the longest read length. Details of three sequencing system are list in Tables 1(a), 1(b), and 1(c).

COMPACT PGM SEQUENCERS

Ion Personal Genome Machine (PGM) and MiSeq were launched by Ion Torrent and Illumina. They are both small in size and feature fast turnover rates but limited data throughput. They are targeted to clinical applications and small labs.

Ion PGM from Ion Torrent

Ion PGM was released by Ion Torrent at the end of 2010. PGM uses semiconductor sequencing technology. When a nucleotide is incorporated into the DNA molecules by the polymerase, a proton is released. By detecting the change in pH, PGM recognized whether the nucleotide is added or not. Each time the chip was flooded with one nucleotide after another, if it is not the correct nucleotide, no voltage will be found; if there is 2 nucleotides added, there is double voltage detected [15]. PGM is the first commercial sequencing machine that does not require fluorescence and camera scanning, resulting in higher speed, lower cost, and smaller instrument size. Currently, it enables 200 bp reads in 2 hours and the sample preparation time is less than 6 hours for 8 samples in parallel.

An exemplary application of the Ion Torrent PGM sequencer is the identification of microbial pathogens. In May and June of 2011, an ongoing outbreak of exceptionally virulent Shiga-toxin- (Stx) producing Escherichia coli O104:H4 centered in Germany [16, 17], there were more than 3000 people infected. The whole genome sequencing on Ion Torrent PGM sequencer and HiSeq 2000 helped the scientists to identify the type of E. coli which would directly apply the clue to find the antibiotic resistance. The strain appeared to be a hybrid of two E. coli strains—entero aggregative E. coli and entero hemorrhagic E. coli—which may help explain why it has been particularly pathogenic. From the sequencing result of E. coli TY2482 [18], PGM shows the potential of having a fast, but limited throughput sequencer when there is an outbreak of new disease. In order to study the sequencing quality, mapping rate, and GC depth distribution of Ion Torrent and compare with HiSeq 2000, a high GC Rhodobacter sample with high GC content (66%) and 4.2 Mb genome was sequenced in these two different sequencers (Table 2). In another experiment, E. coli K12 DH10B (NC_010473.1) with GC 50.78% was sequenced by Ion Torrent for analysis of quality value, read length, position accuracies, and GC distribution (Figure 1).

Table 2: Comparison in alignment between Ion Torrent and HiSeq 2000

	Ion Torrent[a]	**HiSeq 2000**[b]
Total reads num	165518	205683
Total bases num	18574086	18511470
Max read length	201	90
Min read length	15	90
Map reads num	157258	157511
Map rate	95%	76.57%
Covered rate	96.50%	93.11%
Total map length	15800258	14176420
Total mismatch base	53475	142425
Total insertion base	109550	1397
Total insertion num	95740	1332
Total deletion base	152495	431
Total deletion num	139264	238
Ave mismatch rate	0.338%	1.004%
Ave insertion rate	0.693%	0.009%
Ave deletion rate	0.965%	0.003%

[a]: use TMAP to align; [b]: use SOAP2 to align.

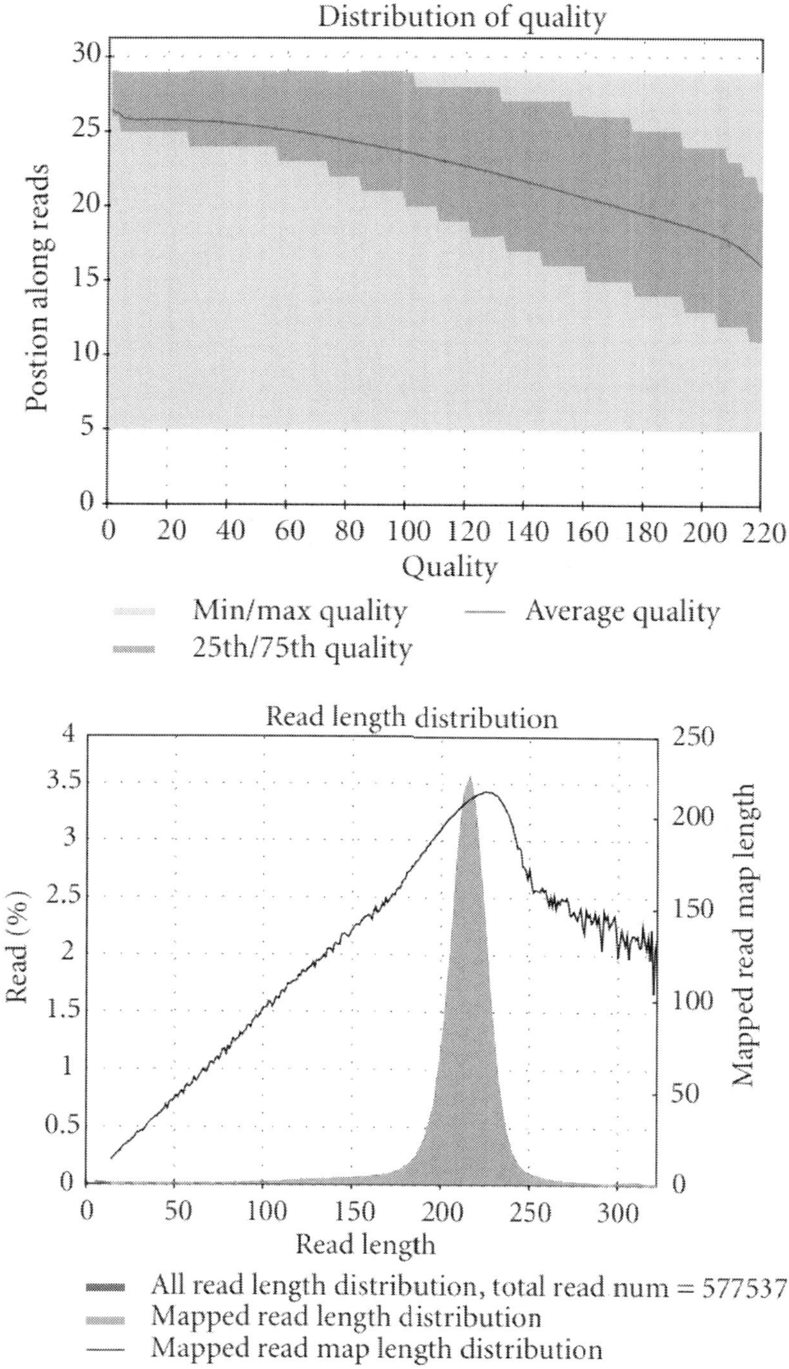
Distribution of quality
Postion along reads
Quality
0
5
10
15
20
25
30
0 20 40 60 80 100 120 140 160 180 200 220
Min/max quality
Average quality
25th/75th quality
Read length distribution
Read (%)
Mapped read map length
Read length
0 0.5 1 1.5 2 2.5 3 3.5 4
0 50 100 150 200 250
0 50 100 150 200 250 300
All read length distribution, total read num = 577537
Mapped read length distribution
Mapped read map length distribution

(a)

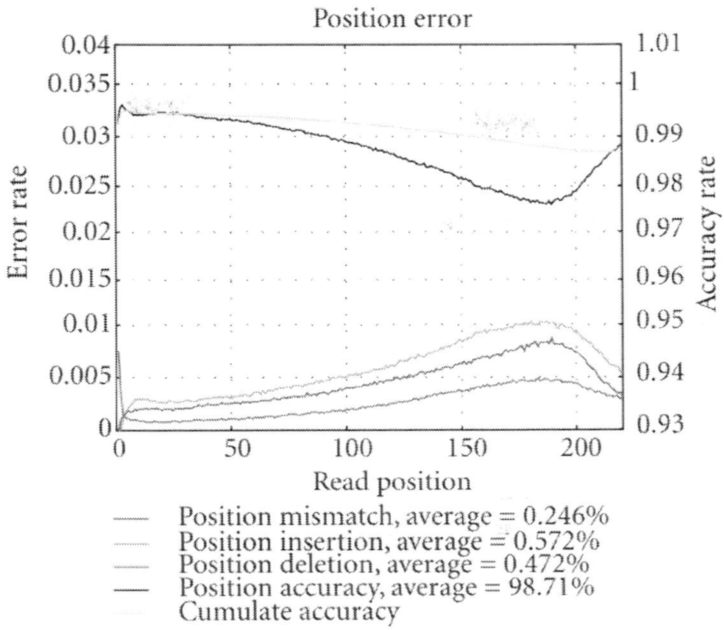
Position error
0.04
0.035
0.03
0.025
0.02
0.015
0.01
0.005
0
Error rate
1.01
1
0.99
0.98
0.97
0.96
0.95
0.94
0.93
Accuracy rate
0
50
100
150
200
Read position
Position mismatch, average = 0.246%
Position insertion, average = 0.572%
Position deletion, average = 0.472%
Position accuracy, average = 98.71%
Cumulate accuracy

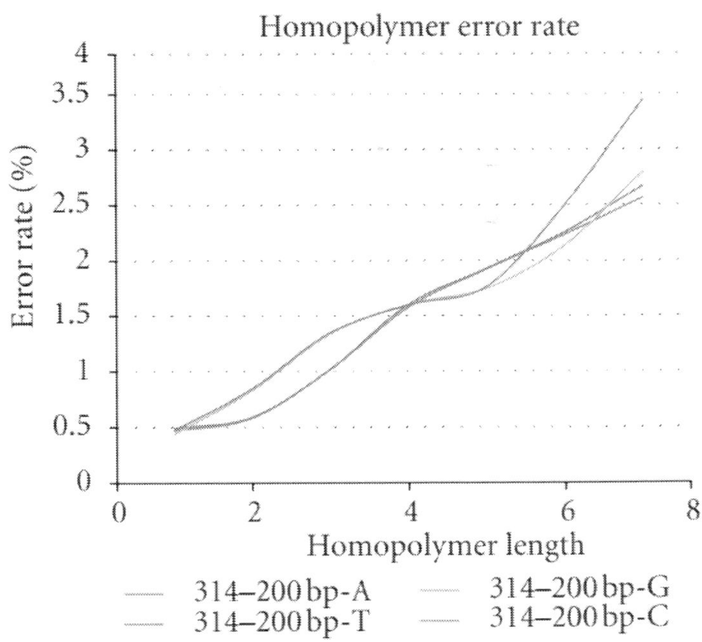
Homopolymer error rate
4
3.5
3
2.5
2
1.5
1
0.5
0
Error rate (%)
0
2
4
6
8
Homopolymer length
314–200 bp-A
314–200 bp-T
314–200 bp-G
314–200 bp-C

(b)

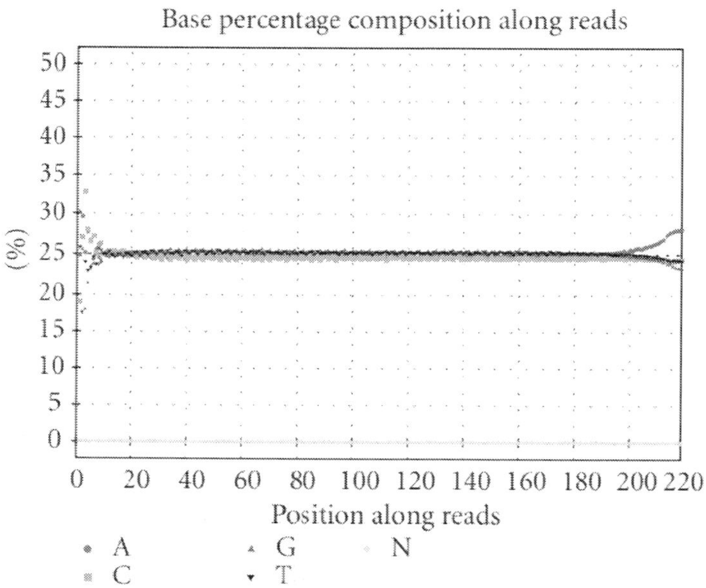

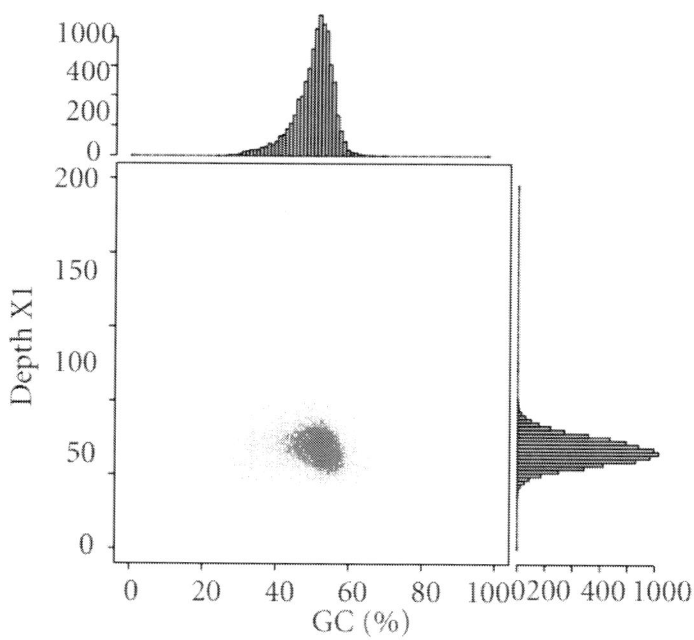

(c)

Figure 1: Ion Torrent sequencing quality. E. coli K12 DH10B (NC_010473.1) with GC 50.78% was used for this experiment. (a) is 314–200 bp from Ion Torrent. The left figure is quality value: pink range represents quality minimum and maximum values each position has. Green area represents the top and bottom quarter (1/4) reads of quality. Red line represents the average quality value in the position. The right figure is read length analysis: colored histogram represents the real read length. The black line represents the mapped length, and because it allows 3′ soft clipping, the length is different from the real read length. (b) is accuracy analysis. In each position, accuracy type including mismatch, insertion, and deletion is shown on the left y-axis. The average accuracy is shown the right y-axis. Accuracy of 200 bp sequencing could reach 99%. (c) is base composition along reads (left) and GC distribution analysis (right). The left figure is base composition in each position of reads. Base line splits after about 95 cycles indicating an inaccurate sequencing. The right one uses 500 bp window and the GC distribution is quite even. The data using high GC samples also indicates a good performance in Ion Torrent (data not shown).

Sequencing Quality

The quality of Ion Torrent is more stable, while the quality of HiSeq 2000 decreases noticeably after 50 cycles, which may be caused by the decay of fluorescent signal with increasing the read length (shown in Figure 1).

Mapping

The insert size of library of Rhodobacter was 350 bp, and 0.5 Gb data was obtained from HiSeq. The sequencing depth was over 100x, and the contig and scaffold N50 were 39530 bp and 194344 bp, respectively. Based on the assembly result, we used 33 Mb which is obtained from ion torrent with 314 chip to analyze the map rate. The alignment comparison is Table 2.

The map rate of Ion Torrent is higher than HiSeq 2000, but it is incomparable because of the different alignment methods used in different sequencers. Besides the significant difference on data including mismatch rate, insertion rate, and deletion rate, HiSeq 2000 and Ion Torrent were still incomparable because of the different sequencing principles. For example, the polynucleotide site could not be indentified easily in Ion Torrent. But it is shown that Ion Torrent has a stable quality along sequencing reads and a good performance on mismatch accuracies, but rather a bias in detection of indels. Different types of accuracy are analyzed and shown in Figure 1.

GC Depth Distribution

The GC depth distribution is better in Ion Torrent from Figure 1. In Ion Torrent, the sequencing depth is similar while the GC content is from 63% to 73%. However in HiSeq 2000, the average sequencing depth is 4x when the GC content is 60%, while it is 3x with 70% GC content.

Ion Torrent has already released Ion 314 and 316 and planned to launch Ion 318 chips in late 2011. The chips are different in the number of wells resulting in higher production within the same sequencing time. The Ion 318 chip enables the production of >1 Gb data in 2 hours. Read length is expected to increase to >400 bp in 2012.

MiSeq from Illumina

MiSeq which still uses SBS technology was launched by Illumina. It integrates the functions of cluster generation, SBS, and data analysis in a single instrument and can go from sample to answer (analyzed data) within a single day (as few as 8 hours). The Nextera, TruSeq, and Illumina's reversible terminator-based sequencing by synthesis chemistry was used in this innovative engineering. The highest integrity data and broader range of application, including amplicon sequencing, clone checking, ChIP-Seq, and small genome sequencing, are the outstanding parts of MiSeq. It is also flexible to perform single 36 bp reads (120 MB output) up to 2 × 150 paired-end reads (1–1.5 GB output) in MiSeq. Due to its significant improvement in read length, the resulting data performs better in contig assembly compared with HiSeq (data not shown). The related sequencing result of MiSeq is shown in Table 3. We also compared PGM with MiSeq in Table 4.

Table 3: MiSeq 150PE data

Sample	**GC**	**Q20**	**Q30**
Human HPV	33.57; 33.62	98.26; 95.52	93.64; 88.52
Bacteria	61.33; 61.43	90.84; 83.86	78.46; 69.04

(1) The data in the table includes both read 1 and read 2 from paired-end sequencing.

(2) GC represents the GC content of libraries.

(3) Q20 value is the average Q20 of all bases in a read, which represents the ratio of bases with probability of containing no more than one error in 100 bases. Q30 value is the average Q30 of all bases in a read, which represents the ratio of bases with probability of containing no more than one error in 1,000 bases.

Table 4: The comparison between PGM and MiSeq

	PGM	**MiSeq**
Output	10 MB–100 MB	120 MB–1.5 GB
Read length	~200 bp	Up to 2 × 150 bp
Sequencing time	2 hours for 1 × 200 bp	3 hours for 1 × 36 single read
		27 hours for 2 × 150 bp pair end read
Sample preparation time	8 samples in parallel, less than 6 hrs	As fast as 2 hrs, with 15 minutes hand on time
Sequencing method	semiconductor technology with a simple sequencing chemistry	Sequencing by synthesis (SBS)
Potential for development	Various parameters (read length, cycle time, accuracy, etc.)	Limited factors, major concentrate in flowcell surface size, insert sizes, and how to pack cluster in tighter
Input amount	μg	Ng (Nextera)
Data analysis	Off instrument	On instrument

Complete Genomics

Complete genomics has its own sequencer based on Polonator G.007, which is ligation-based sequencer. The owner of Polonator G.007, Dover, collaborated with the Church Laboratory of Harvard Medical School, which is the same team as SOLiD system, and introduced this cheap open system. The Polonator could combine a high-performance instrument at very low price and the freely downloadable, open-source software and protocols in this sequencing system. The Polonator G.007 is ligation detection sequencing, which decodes the base by the single-base probe in nonanucleotides (nonamers), not by dual-base coding [19]. The fluorophore-tagged nonamers will be degenerated by selectively ligate onto a series of anchor primers, whose four components are labeled with one of four fluorophores with the help of T4 DNA ligase, which correspond to the base type at the query position. In the ligation progress, T4 DNA ligase is particularly sensitive to mismatches on 3′-side of the gap which is benefit to improve the accuracy of sequencing. After imaging, the Polonator chemically strips the array of annealed primer-fluorescent probe complex; the anchor primer is replaced and the new mixture are fluorescently tagged nonamers is introduced to sequence the adjacent base [20]. There are two updates compared with Polonator G.007, DNA nanoball (DNB) arrays, and combinatorial probe-anchor ligation (cPAL). Compared with DNA cluster or microsphere, DNA nanoball arrays obtain higher density of DNA cluster on the surface of a silicon chip. As the seven 5-base segments are discontinuous, so the system of hybridization-ligation-detection cycle has higher fault-tolerant ability compared with SOLiD. Complete genomics claim to have 99.999% accuracy with 40x depth and could analyze SNP, indel, and CNV with price 5500$–9500$. But Illumina reported a better performance of HiSeq 2000 use only 30x data (Illumina Genome Network). Recently some researchers compared CG's human genome sequencing data with Illumina system [21], and there are notable differences in detecting SNVs, indels, and system-specific detections in variants.

The Third Generation Sequencer

While the increasing usage and new modification in next generation sequencing, the third generation sequencing is coming out with new insight in the sequencing. Third-generation sequencing has two main characteristics. First, PCR is not needed before sequencing, which shortens DNA preparation time for sequencing. Second, the signal is captured in real time, which means that the signal, no matter whether it is fluorescent (Pacbio) or electric current (Nanopore), is monitored during the enzymatic reaction of adding nucleotide in the complementary strand.

Single-molecule real-time (SMRT) is the third-generation sequencing method developed by Pacific Bioscience (Menlo Park, CA, USA), which made use of modified enzyme and direct observation of the enzymatic reaction in real time. SMRT cell consists of millions of zero-mode waveguides (ZMWs), embedded with only one set of enzymes and DNA template that can be detected during the whole process. During the reaction, the enzyme will incorporate the nucleotide into the complementary strand

and cleave off the fluorescent dye previously linked with the nucleotide. Then the camera inside the machine will capture signal in a movie format in real-time observation [19]. This will give out not only the fluorescent signal but also the signal difference along time, which may be useful for the prediction of structural variance in the sequence, especially useful in epigenetic studies such as DNA methlyation [22].

Comparing to second generation, PacBio RS (the first sequencer launched by PacBio) has several advantages. First the sample preparation is very fast; it takes 4 to 6 hours instead of days. Also it does not need PCR step in the preparation step, which reduces bias and error caused by PCR. Second, the turnover rate is quite fast; runs are finished within a day. Third, the average read length is 1300 bp, which is longer than that of any second-generation sequencing technology. Although the throughput of the PacBioRS is lower than second-generation sequencer, this technology is quite useful for clinical laboratories, especially for microbiology research. A paper has been published using PacBio RS on the Haitian cholera outbreak [19].

We have run a de novo assembly of DNA fosmid sample from Oyster with PacBio RS in standard sequencing mode (using LPR chemistry and SMRTcells instead of the new version FCR chemistry and SMRTcells). An SMRT belt template with mean insert size of 7500 kb is made and run in one SMRT cell and a 120-minute movie is taken. After Post-QC filter, 22,373,400 bp reads in 6754 reads (average 2,566 bp) were sequenced with the average Read Score of 0.819. The Coverage is 324x with mean read score of 0.861 and high accuracy (~99.95). The result is exhibited in Figure 2.

	Prefilter	Post-QC filter*
Number of bases	84, 110, 272 bp	22, 373, 400 bp
Number of reads	46, 861	6, 754
Mean read length	513 bp	2, 566 bp
Mean read score	0.144	0.819

* MinRL = 50, MinRS = 0.75

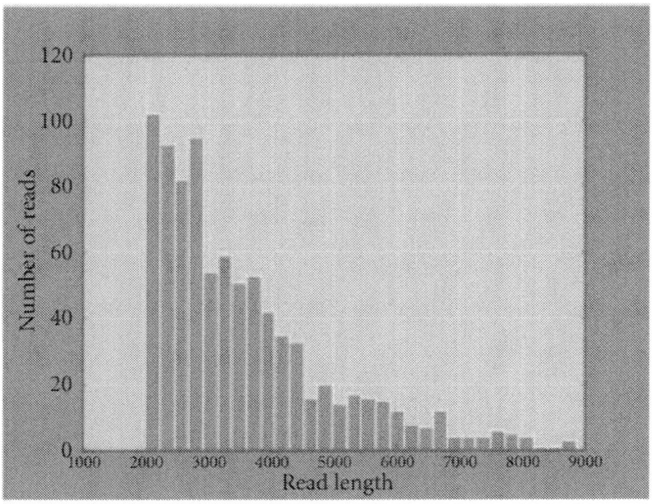
Number of reads
120
100
80
60
40
20
0
1000
2000
3000
4000
5000
6000
7000
8000
9000
Read length

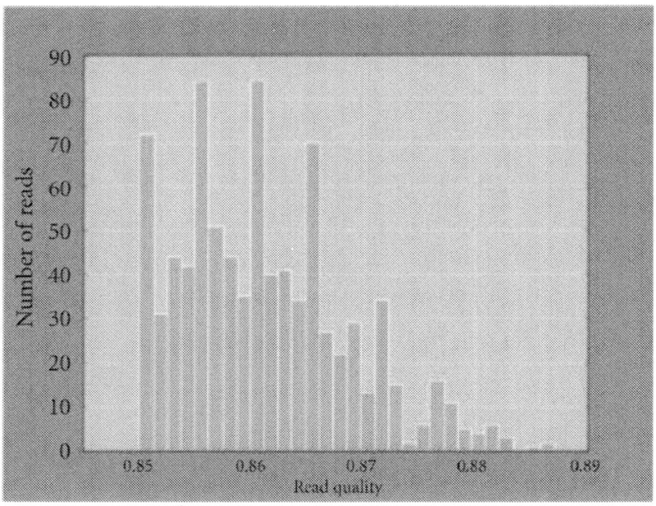
Number of reads
90
80
70
60
50
40
30
20
10
0
0.85
0.86
0.87
0.88
0.89
Read quality

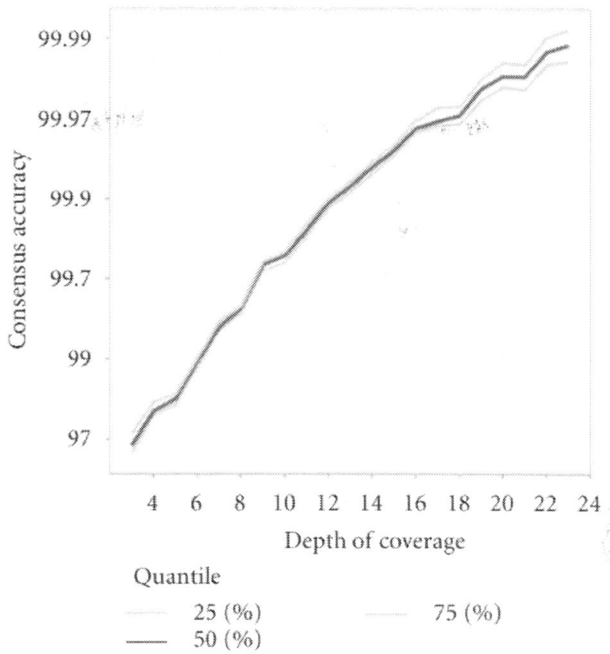

Figure 2: Sequencing of a fosmid DNA using Pacific Biosciences sequencer. With coverage, the accuracy could be above 97%. The figure was constructed by BGI's own data.

Nanopore sequencing is another method of the third generation sequencing. Nanopore is a tiny biopore with diameter in nanoscale [23], which can be found in protein channel embedded on lipid bilayer which facilitates ion exchange. Because of the biological role of nanopore, any particle movement can disrupt the voltage across the channel. The core concept of nanopore sequencing involves putting a thread of single-stranded DNA across α-haemolysin (αHL) pore. αHL, a 33 kD protein isolated from Staphylococcus aureus [20], undergoes self-assembly to form a heptameric transmembrane channel [23]. It can tolerate extraordinary voltage up to 100 mV with current 100 pA [20]. This unique property supports its role as building block of nanopore. In nanopore sequencing, an ionic flow is applied continuously. Current disruption is simply detected by standard electrophysiological technique. Readout is relied on the size difference between all deoxyribonucleoside monophosphate (dNMP). Thus, for given dNMP, characteristic current modulation is shown for discrimination. Ionic current is resumed after trapped nucleotide entirely squeezing out. Nanopore sequencing possesses a number of fruitful advantages over existing commercialized next-generation sequencing technologies. Firstly, it potentially reaches long read length >5 kbp with speed 1 bp/ns [19]. Moreover, detection of bases is fluorescent tag-free. Thirdly, except the use of exonuclease for holding up ssDNA and nucleotide cleavage [24], involvement of enzyme is remarkably obviated in nanopore sequencing [22]. This implies that nanopore sequencing is less sensitive to temperature throughout the sequencing reaction and reliable outcome can be maintained. Fourthly, instead of sequencing DNA during polymerization, single DNA strands are sequenced through nanopore by means of DNA strand depolymerization. Hence, hand-on time for sample preparation such as cloning and amplification steps can be shortened significantly.

DISCUSSION OF NGS APPLICATIONS

Fast progress in DNA sequencing technology has made for a substantial reduction in costs and a substantial increase in throughput and accuracy. With more and more organisms being sequenced, a flood of genetic data is inundating the world every day. Progress in genomics has been moving steadily forward due to a revolution in sequencing technology. Additionally, other of types-large scale studies in exomics, metagenomics, epigenomics, and transcriptomics all become reality. Not only do these studies provide the knowledge for basic research, but also they afford immediate application benefits. Scientists across many fields are utilizing these data for the development of better-thriving crops and crop yields and livestock and improved diagnostics, prognostics, and therapies for cancer and other complex diseases.

BGI is on the cutting edge of translating genomics research into molecular breeding and disease association studies with belief that agriculture, medicine, drug development, and clinical treatment will eventually enter a new stage for more detailed understanding of the genetic components of all the organisms. BGI is primarily focused on three projects.

1) The Million Species/Varieties Genomes Project, aims to sequence a million economically and scientifically important plants, animals, and model organisms, including different breeds, varieties,

and strains. This project is best represented by our sequencing of the genomes of the Giant panda, potato, macaca, and others, along with multiple resequencing projects.

2) The Million Human Genomes Project focuses on large-scale population and association studies that use whole-genome or whole-exome sequencing strategies.

3) The Million Eco-System Genomes Project has the objective of sequencing the metagenome and cultured microbiome of several different environments, including microenvironments within the human body [25]. Together they are called 3 M project.

In the following part, each of the following aspects of applications including de novo sequencing, mate-pair, whole genome or target-region resequencing, small RNA, transcriptome, RNA seq, epigenomics, and metagenomics, is briefly summarized.

In DNA de novo sequencing, the library with insert size below 800 bp is defined as DNA short fragment library, and it is usually applied in de novo and resequencing research. Skovgaard et al. [26] have applied a combination method of WGS (whole-genome sequencing) and genome copy number analysis to identify the mutations which could suppress the growth deficiency imposed by excessive initiations from the E. coli origin of replication, oriC.

Mate-pair library sequencing is significant beneficial for de novo sequencing, because the method could decrease gap region and extend scaffold length. Reinhardt et al. [27] developed a novel method for de novogenome assembly by analyzing sequencing data from high-throughput short read sequencing technology. They assembled genomes into large scaffolds at a fraction of the traditional cost and without using reference sequence. The assembly of one sample yielded an N50 scaffold size of 531,821 bp with >75% of the predicted genome covered by scaffolds over 100,000 bp.

Whole genome resequencing sequenced the complete DNA sequence of an organism's genome including the whole chromosomal DNA at a single time and alignment with the reference sequence. Mills et al. [28] constructed a map of unbalanced SVs (genomic structural variants) based on whole genome DNA sequencing data from 185 human genomes with SOLiD platform; the map encompassed 22,025 deletions and 6,000 additional SVs, including insertions and tandem duplications [28]. Most SVs (53%) were mapped to nucleotide resolution, which facilitated analyzing their origin and functional impact [28].

The whole genome resequencing is an effective way to study the functional gene, but the high cost and massive data are the main problem for most researchers. Target region sequencing is a solution to solve it. Microarray capture is a popular way of target region sequencing, which uses hybridization to arrays containing synthetic oligo-nucleotides matching the target DNA sequencing. Gnirke et al. [29] developed a captured method that uses an RNA "baits" to capture target DNA fragments from the "pond" and then uses the Illumina platform to read out the sequence. About 90% of uniquely aligning bases fell on or near bait sequence; up to 50% lay on exons proper [29].

Fehniger et al. used two platforms, Illumina GA and ABI SOLiD, to define the miRNA transcriptomes of resting and cytokine-activated primary murine NK (natural killer) cells [30]. The identified 302 known and 21 novel mature miRNAs were analyzed by unique bioinformatics pipeline from small RNA libraries of NK cell. These miRNAs are overexpressed in broad range and exhibit isomiR complexity, and a subset is differentially expressed following cytokine activation, which were the clue to identify the identification of miRNAs by the Illumina GA and SOLiD instruments [30].

The transcriptome is the set of all RNA molecules, including mRNA, rRNA, tRNA, and other noncoding RNA produced in one or a population of cells. In these years, next-generation sequencing technology is used to study the transcriptome compares with DNA microarray technology in the past. The S. mediterranea transcriptome could be sequenced by an efficient sequencing strategy which designed by Adamidi et al. [31]. The catalog of assembled transcripts and the identified peptides in this study dramatically expand and refine planarian gene annotation, which is demonstrated by validation of several previously unknown transcripts with stem cell-dependent expression patterns.

RNA-seq is a new method in RNA sequencing to study mRNA expression. It is similar to transcriptome sequencing in sample preparation, except the enzyme. In order to estimate the technical variance, Marioni et al. [32] analyzed a kidney RNA samples on both Illumina platform and Affymetrix arrays. The additional analyses such as low-expressed genes, alternative splice variants, and novel transcripts were found on Illumina platform. Bradford et al. [33] compared the data of RNA-seq library on the SOLiD platform and Affymetrix Exon 1.0ST arrays and found a high degree of correspondence between the two platforms in terms of exon-level fold changes and detection. And the greatest detection correspondence was seen when the background error rate is extremely low in RNA-seq. The difference between RNA-seq and transcriptome on SOLiD is not so obvious as Illumina.

There are two kinds of application of epigenetic, Chromatin immunoprecipitation and methylation analysis. Chromatin immunoprecipitation (ChIP) is an immunoprecipitation technique which is used to study the interaction between protein and DNA in a cell, and the histone modifies would be found by the specific location in genome. Based on next-generation sequencing technology, Johnson et al. [34] developed a large-scale chromatin immunoprecipitation assay to identify motif, especially noncanonical NRSF-binding motif. The data displays sharp resolution of binding position (±50 bp), which is important to infer new candidate interaction for the high sensitivity and specificity (ROC (receiver operator characteristic) area ≥0.96) and statistical confidence ($P < 10–4$). Another important application in epigenetic is DNA methylation analysis. DNA methylation exists typically in vertebrates at CpG sites; the methylation caused the conversion of the cytosine to 5-methylcytosine. Chung presented a whole methylome sequencing to study the difference between two kinds of bisulfite conversion methods (in solution versus in gel) by SOLiD platform [35].

The world class genome projects include the 1000 genome project, and the human ENCODE project, the human Microbiome (HMP) project, to name a few. BGI takes an active role in these and many more ongoing projects like 1000 Animal and Plant Genome project, the MetaHIT project, Yanhuang project, LUCAMP (Diabetes-associated Genes and Variations Study), ICGC (international cancer genome project), Ancient human genome, 1000 Mendelian Disorders Project, Genome 10 K Project, and so forth [25]. These internationally collaborated genome projects greatly enhanced genomics study and applications in healthcare and other fields.

To manage multiple projects including large and complex ones with up to tens of thousands of samples, a superior and sophisticated project management system is required handling information processing from the very beginning of sample labeling and storage to library construction, multiplexing, sequencing, and informatics analysis. Research-oriented bioinformatics analysis and followup experiment processed are not included. Although automation techniques' adoption has greatly simplified bioexperiment human interferences, all other procedures carried out by human power have to be managed. BGI has developed BMS system and Cloud service for efficient information exchange and project management. The behavior management mainly follows Japan 5S onsite model. Additionally, BGI has passed ISO9001 and CSPro (authorized by Illumina) QC system and is currently taking (Clinical Laboratory Improvement Amendments) CLIA and (American Society for Histocompatibility and Immunogenetics) AShI tests. Quick, standard, and open reflection system guarantees an efficient troubleshooting pathway and high performance, for example, instrument design failure of Truseq v3 flowcell resulting in bubble appearance (which is defined as "bottom-middle-swatch" phenomenon by Illumina) and random *N* in reads. This potentially hazards sequencing quality, GC composition as well as throughput. It not only effects a small area where the bubble locates resulting in reading *N* but also effects the focus of the place nearby, including the whole swatch, and the adjacent swatch. Filtering parameters have to be determined to ensure quality raw data for bioinformatics processing. Lead by the NGS tech group, joint meetings were called for analyzing and troubleshooting this problem, to discuss strategies to best minimize effect in terms of cost and project time, to construct communication channel, to statistically summarize compensation, in order to provide best project management strategies in this time. Some reagent QC examples are summaried in Liu et al. [36].

BGI is establishing their cloud services. Combined with advanced NGS technologies with multiple choices, a plug-and-run informatics service is handy and affordable. A series of softwares are available including BLAST, SOAP, and SOAP SNP for sequence alignment and pipelines for RNAseq data. Also SNP calling programs such as Hecate and Gaea are about to be released. Big-data studies from the whole spectrum of life and biomedical sciences now can be shared and published on a new journal GigaSicence cofounded by BGI and Biomed Central. It has a novel publication format: each piece of data links to a standard manuscript publication with an extensive database which hosts all associated data, data analysis tools, and cloud-computing resources. The scope covers not just omic type data and the fields of high-throughput biology currently serviced by large public repositories but also the growing range of more difficult-to-access data, such as imaging, neuroscience, ecology, cohort data, systems biology, and other new types of large-scale sharable data.

REFERENCES

1. G. M. Church and W. Gilbert, "Genomic sequencing," Proceedings of the National Academy of Sciences of the United States of America, vol. 81, no. 7, pp. 1991–1995, 1984.
2. http://en.wikipedia.org/wiki/DNA_sequencing/.
3. F. S. Collins, M. Morgan, and A. Patrinos, "The Human Genome Project: lessons from large-scale biology," Science, vol. 300, no. 5617, pp. 286–290, 2003. · ·
4. http://genomics.xprize.org/.
5. http://my454.com/products/technology.asp.
6. J. Berka, Y. J. Chen, J. H. Leamon et al., "Bead emulsion nucleic acid amplification," U.S. Patent Application, 2005.
7. T. Foehlich et al., "High-throughput nucleic acid analysis," U.S. Patent, 2010.
8. http://www.pyrosequencing.com/DynPage.aspx.
9. http://www.roche-applied-science.com/.
10. E. R. Mardis, "The impact of next-generation sequencing technology on genetics," Trends in Genetics, vol. 24, no. 3, pp. 133–141, 2008. · ·
11. S. M. Huse, J. A. Huber, H. G. Morrison, M. L. Sogin, and D. M. Welch, "Accuracy and quality of massively parallel DNA pyrosequencing," Genome Biology, vol. 8, no. 7, article R143, 2007. · ·
12. "The new GS junior sequencer," http://www.gsjunior.com/instrument-workflow.php.
13. "SOLiD system accuray," http://www.appliedbiosystems.com/absite/us/en/home/applications-technologies/solid-next-generation-

sequencing.html.

14. http://www.tecan.com/platform/apps/product/index.asp?MenuID=3465&ID=7191&Menu=1&Item=33.52.2.
15. B. A. Flusberg, D. R. Webster, J. H. Lee et al., "Direct detection of DNA methylation during single-molecule, real-time sequencing," Nature Methods, vol. 7, no. 6, pp. 461–465, 2010. · ·
16. A. Mellmann, D. Harmsen, C. A. Cummings et al., "Prospective genomic characterization of the german enterohemorrhagic Escherichia coli O104:H4 outbreak by rapid next generation sequencing technology," PLoS ONE, vol. 6, no. 7, Article ID e22751, 2011.
17. H. Rohde, J. Qin, Y. Cui et al., "Open-source genomic analysis of Shiga-toxin-producing E. coli O104:H4," New England Journal of Medicine, vol. 365, no. 8, pp. 718–724, 2011.
18. C. S. Chin, J. Sorenson, J. B. Harris et al., "The origin of the Haitian cholera outbreak strain," New England Journal of Medicine, vol. 364, no. 1, pp. 33–42, 2011. · ·
19. W. Timp, U. M. Mirsaidov, D. Wang, J. Comer, A. Aksimentiev, and G. Timp, "Nanopore sequencing: electrical measurements of the code of life," IEEE Transactions on Nanotechnology, vol. 9, no. 3, pp. 281–294, 2010. · ·
20. D. W. Deamer and M. Akeson, "Nanopores and nucleic acids: prospects for ultrarapid sequencing," Trends in Biotechnology, vol. 18, no. 4, pp. 147–151, 2000. · ·
21. "Performance comparison of whole-genome sequencing systems," Nature Biotechnology, vol. 30, pp. 78–82, 2012.
22. D. Branton, D. W. Deamer, A. Marziali et al., "The potential and challenges of nanopore sequencing," Nature Biotechnology, vol. 26, no. 10, pp. 1146–1153, 2008.
23. L. Song, M. R. Hobaugh, C. Shustak, S. Cheley, H. Bayley, and J. E. Gouaux, "Structure of staphylococcal α-hemolysin, a heptameric transmembrane pore," Science, vol. 274, no. 5294, pp. 1859–1866, 1996. · ·
24. J. Clarke, H. C. Wu, L. Jayasinghe, A. Patel, S. Reid, and H. Bayley, "Continuous base identification for single-molecule nanopore DNA

sequencing," Nature Nanotechnology, vol. 4, no. 4, pp. 265–270, 2009. · ·

25. Website of BGI, http://www.genomics.org.cn.
26. O. Skovgaard, M. Bak, A. Løbner-Olesen et al., "Genome-wide detection of chromosomal rearrangements, indels, and mutations in circular chromosomes by short read sequencing," Genome Research, vol. 21, no. 8, pp. 1388–1393, 2011.
27. J. A. Reinhardt, D. A. Baltrus, M. T. Nishimura, W. R. Jeck, C. D. Jones, and J. L. Dangl, "De novo assembly using low-coverage short read sequence data from the rice pathogen Pseudomonas syringae pv. oryzae," Genome Research, vol. 19, no. 2, pp. 294–305, 2009. · ·
28. R. E. Mills, K. Walter, C. Stewart et al., "Mapping copy number variation by population-scale genome sequencing," Nature, vol. 470, no. 7332, pp. 59–65, 2011. · ·
29. A. Gnirke, A. Melnikov, J. Maguire et al., "Solution hybrid selection with ultra-long oligonucleotides for massively parallel targeted sequencing," Nature Biotechnology, vol. 27, no. 2, pp. 182–189, 2009. · ·
30. T. A. Fehniger, T. Wylie, E. Germino et al., "Next-generation sequencing identifies the natural killer cell microRNA transcriptome," Genome Research, vol. 20, no. 11, pp. 1590–1604, 2010. · ·
31. C. Adamidi, Y. Wang, D. Gruen et al., "De novo assembly and validation of planaria transcriptome by massive parallel sequencing and shotgun proteomics," Genome Research, vol. 21, no. 7, pp. 1193–1200, 2011. · ·
32. J. C. Marioni, C. E. Mason, S. M. Mane, M. Stephens, and Y. Gilad, "RNA-seq: an assessment of technical reproducibility and comparison with gene expression arrays," Genome Research, vol. 18, no. 9, pp. 1509–1517, 2008. · ·
33. J. R. Bradford, Y. Hey, T. Yates, Y. Li, S. D. Pepper, and C. J. Miller, "A comparison of massively parallel nucleotide sequencing with oligonucleotide microarrays for global transcription profiling," BMC Genomics, vol. 11, no. 1, article 282, 2010. · ·
34. D. S. Johnson, A. Mortazavi, R. M. Myers, and B. Wold, "Genome-wide mapping of in vivo protein-DNA interactions," Science, vol.

316, no. 5830, pp. 1497–1502, 2007. · ·

35. H. Gu, Z. D. Smith, C. Bock, P. Boyle, A. Gnirke, and A. Meissner, "Preparation of reduced representation bisulfite sequencing libraries for genome-scale DNA methylation profiling," Nature Protocols, vol. 6, no. 4, pp. 468–481, 2011. · ·
36. L. Liu, N. Hu, B. Wang et al., "A brief utilization report on the Illumina HiSeq 2000 sequencer," Mycology, vol. 2, no. 3, pp. 169–191, 2011.

CHAPTER 3

NEXT GENERATION DNA SEQUENCING (II): TECHNIQUES, APPLICATIONS

Wilhelm J Ansorge

Ecole Polytechnique Federal Lausanne, EPFL, Lausanne, Switzerland

ABSTRACT

Development and applications of DNA sequencing techniques in the last years have exceeded by far the expectations. Historical development of the field from the start up to the initial Next Generation systems, technical principles of the platforms existing 6 years ago and many applications, have been described in the previous reviews (e.g. 1,2). In this updated review is discussed the status of the technology in the second half of the year 2015, both of the novel commercially available platforms, and of the systems in development. Mentioned are techniques with potential for future DNA sequencing techniques. Discussed are some of the not solved challenges for the platforms, the price to performance ratio, complexity of sample preparation, and detection of structural variations in genome. The aims for the technology will be to lower the cost of the equipment and biochemicals involved, increasing simultaneously the reproducibility, reliability and simplicity of the techniques and protocols in operation, of importance for diagnostics and applications in the clinics. During the last few years have been rapidly developing technologies and methods that permit analysis of the genome and transcriptome of a single cell. The first observations suggest that both genomic and transcriptomic heterogeneities within an organism are more common than expected, during normal development and disease. Recent public support for precision medicine, for novel therapeutic approaches and efforts for improvements in the healthcare, will be a motivation for further innovations and developments in the field.

Keywords

DNA sequencing, Transcriptome; Single cell, Nanopore

INTRODUCTION

From the time of the previous reviews 6 years ago [1,2], DNA sequencing techniques and its applications in the diagnostics field have made significant progress. The innovations are reflected particularly in higher throughput and lower cost per base, e.g. on Illumina and Ion Torrent systems, or the single-molecule real time sequencing approaches of Pacific Bioscience and Oxford Nanopore. The human genome sequence has complemented the understanding of biology, human diversity, and disease. The transition from the first human genome sequences to the personal genomes and genomic medicine has been made possible only because of the advances in DNA sequencing technologies over the past 12 years. NGS utilizes massively parallel sequencing to generate up to several gigabases of sequence information per day, making possible new applications that were in the past difficult to perform. The technique has spread to almost all fields concerning DNA, to such various areas as food control, screening of the purity of vaccines and other medicaments (where it can detect not only traces of the active viral components but also the presence of nonactive viral particles), evolution of organisms, analysis of fetus DNA from the mother's blood, forensic field, and others.

Historical development of the DNA sequencing field from the start up to the initial next generation systems, as well as the technical principles of some of the platforms existing already 6 years ago, and many applications, have been described in the previous reviews (e.g. 1,2). In this updated review is discussed the status of the technology in the second half of the year 2015, both of the commercially available platforms, and of the systems in the development (announced to be in an advanced phase close to an introduction on the market). Discussed are some of the not solved challenges for the platforms, the price to performance ratio, simplicity and reliability of operation, of importance for diagnostics and applications in the clinics. Mentioned are also some techniques reported to be in the initial phase of tests for DNA sequencing.

During the last few years have been rapidly developing technologies and methods that permit analysis of the genome and transcriptome of a single cell. The first observations suggest that both genomic and transcriptomic heterogeneities within an organism are more common than expected, during normal development and disease.

The diagnostics field has already profited from new generations of DNA sequencing technology, sample preparation, single-cell isolation techniques, DNA amplification schemes, and computer data analysis with improving software tools. The applications resulted in better understanding of causes for diseases, in elucidation of reasons why in some persons a drug efficiently treats a disease, while the treatment is not successful in others with the same disease. Recent public support for Precision Medicine, for novel therapeutic approaches and efforts for improvements in the healthcare, will be a motivation for further innovations and developments in the field.

The aims for the technology will be to lower the cost of the equipment and biochemicals involved, increasing simultaneously the reproducibility, reliability and simplicity of the techniques and protocols in operation. The goal to be achieved formulated in [1,2] is still valid, in spite of the great progress in DNA sequencing techniques since then. For genomic sequencing, and for analysis of the ever-more important structural genetic variations in genomes (e.g. copy-number variation CNVs, for chromosomal translocations, inversions, large deletions, insertions, and duplications) it would be a great advantage if sequence read length on the original single DNA molecule could be increased to tens of thousands of bases and more, in a relatively short time. Ideally, the goal would be the sequence determination of a whole chromosome, from a single original DNA molecule, from a single cell.

Commercially Available Platforms

Since the last reviews [1,2], several new companies entered the Next Generation DNA Sequencing (NGS) field (e.g. Oxford Nanopore, Genia, GnuBio, others), with new devices described below.

Some companies limited the distribution of their systems, or are no longer active in the market. The first was Helicos and although pioneer in the field, its revolutionary Single Molecule Detection technique used in the device was not yet mature enough and its production stopped. The technique was recently re-designed, and a prototype of the new device called GenoCare developed and presented in October 2015 by the Direct Genomics company (see section 3.3 below).The SOLiD system platform from Life Sciences did not find decisive favor with users, possibly because of the somewhat complex biochemical sample preparation, high cost and relatively short reading length around 100 bases. The Roche company stopped the production and distribution of the pyrosequencing bioluminescence detection Gene Sequencer, because of its higher not competitive sequencing cost. It was developed originally by the 454 company, as the first system from the Next Generation devices on the market, and was for several years very successfully applied in many projects. After the attempt to take over Illumina, the Roche company has now initiated collaborations with sequencing platforms of Pacific Biosystems and Genia companies, discussed below.

Illumina Platform

The company (www.illumina.com) pushed very successfully its technology developed originally by the Solexa company [1,2], and continuously introduced novel devices both for very high throughput, as well as smaller and more modestly priced versions. At the same time, they improved the reading length from the originally relatively short 50 bases to about 300 bases, improved several times the biochemistry of the sample preparation, and reduced the sequencing cost down to $0.001 per 1000 bases, which would be not expected few years ago.

The sequencing devices offer starts with the MiSeq series, with speed and simplicity suitable for small genomes, amplicons and targeted gene panel sequencing, and it is also foreseen for applications in clinical research. The system has 1 flow cell, output ranges from 0.3 to 15

Gb, number of reads per flow cell 25 million, the run time 5-55 hours, Maximum Read Length 2*300 bp.

The HiSeq series in the middle range is used for sequencing of genomes, exomes and transcriptomes. As example, the HiSeq 2500 device in the rapid mode has 1 or 2 flow cells per run, the output range 10-300 Gb, run time 7-60 hours, results in 300 million reads per flow cell, and maximum read length 2*250 bp.

The HiSeq series with very high throughput is used in the production and population-scale for whole genomes sequencing. As example, the HiSeq 4000 device has 1 or 2 flow cells per run, the output range 125-1500 Gb, run time 1-3.5 days, 2.5 milliards (in the US 2.5 billions) reads per flow cell, and maximum read length 2*150 bp.

The data flow resulting from the large sequencing throughput, and its analysis, requires great efforts on the bioinformatics part. One of the causes may also be the still relatively short reads obtainable with the technique.

There are now many Illumina systems purchased by the laboratories worldwide. It is possible that the global sequencing capacity is sufficiently large for present needs, so the next wave of acquisitions may not grow as fast as the previous one, taking into account also the competing systems being introduced in operation, as described below.

Schema of operation, the numerous applications in analysis of gene expression and genomes, relevant publications, collaborations and operation protocols involving the Illumina systems, are listed at the web site of the company (www.illumina.com). Opinion of users can be found on the Genome web site (www.genomeweb.com).

Ion Torrent-Thermofisher Scientific

Ion Torrent Technology has been a continuation of the original pyrosequencing technique, as explained e.g. in [1,2]. In its original design, the 454 company detected the pyrophosphate group released during the incorporation of a nucleotide in the complementary strand, during the DNA polymerization process. The present Ion Torrent system detects the H+ ions, another product released during the DNA polymerization. The basic principle, the detection of H+ ions, has been licensed from the DNA Electronics company in London. The technology combines semiconductor sequencing technology with biochemistry, enabling the direct translation of chemical information into digital sequence data. This eliminates the need for expensive optics, lasers, and complex sequencing chemistries with fluorescently labeled nucleotides. The result is a sequencing system that is more affordable and cost-effective, faster, scalable and relatively simple to operate.

The Ion PGM System (depending on the Chip used and the application goals) can run up to 5.5 million reads, output reaching to 2 giga bases, reading length up to 400 bases, run time between 2-7 hours.

The Ion Proton System has been designed for high throughput sequencing of exomes, transcriptomes and genomes. With the Ion PI Chip, the throughput is up to 10 Gb, number of reads up to 80 million, read length 200 base, and run time 2-4 hours. It has applications in sequencing of: Genes, Human genomes, De novo sequencing, ChIP technique, whole transcriptomes, exomes, methylation analysis, Gene expression by sequencing, small genomes, and small RNAs.

The NGS devices Ion S5 and Ion S5 XL are being introduced for a simple and rapid workflow for panels, microbes, exomes, transcriptomes, and are applicable both for high and low weekly throughput. The output with Ion 540 Chip is up to 15 Gb (about half with reads 400 bases), number of reads up to 80 million, run time 2.5-5 hours (for reads of 200 or 400 bases). The claim is only 10 ng of low-quality DNA or RNA are needed to generate mutational or gene expression profiles. Target selection is fast and simple using Ion AmpliSeq panels, requiring also only as little as 10 ng of input material. Ion Torrent systems have simplified NGS data analysis with end-to-end bioinformatics solutions, including the Ion Reporter Software and server. Operating software and Plugins are available for analysis of genome sequencing, targeted sequencing, variant analysis and annotation, microbial sequencing, transfer to 3rd party bioinformatics packages, and more.

Because of its affordable price and simplicity of operation, the Ion Torrent systems are widely used worldwide in many projects. Schema of operation, list of publications and collaborations using this system, are obtainable from the company web page (www.thermofisher.com). Opinion of users can be found on the genome website (www.genomeweb.com).

Pacific Biosystems

The Pacific Biosystems has been developing the single-molecule real-time (SMRT) technology for many years. As the technology continuously improved over the years it was successfully attracting an increasing number of users. PacBio Sequencing System may provide very high depth of genetic information through exceptionally long sequencing reads (average >10,000 bp), and the high consensus accuracy.

The basic principle of the technology, explained in (www.pacbio.com), is the detection of the fluorescently labeled nucleotide as it is being incorporated in the complementary strand of the single molecule by the DNA polymerase. In the moment of the incorporation is detected the fluorescent label specific for the base, while the polymerase simultaneously cuts off the label from the nucleotide. The process is repeated for the next labelled nucleotide, and from the order of the 4 different labels detected successively is determined the base sequence. The process takes place in Zero-mode waveguides (ZMW) chambers. Within the ZMWs, single molecule of polymerase is immobilized and fluorescently labeled nucleotide added, so that DNA sequencing can be observed optically and recorded in real time.

In the first PacBio RS II system the SMRT cell contains 150.000 ZMWs, with a throughput of 1 gb per SMRT cell. It has high-performance optics, automated liquid handling, SMRT cells and reagents, instrument control software, run-time from 30 minutes to 6 hours per SMRT cell, run-size from 1 to 16 SMRT cells per run. With this established technology, there are over 850 scientific publications in a variety of research areas, such as human biomedical research, plant and animal genomics, and microbiology. The PacBio RS II has been applied to whole genome sequencing of small genomes, targeted sequencing, complex population analysis, RNA sequencing of targeted transcripts, and microbial epigenetics.

In autumn 2015 Pacific Biosciences launched a new Sequel single molecule sequencing system that offers higher throughput and costs, at about half the price of the original RS II device. PacBio developed the Sequel as part of its collaboration with Roche to develop a clinical grade sequencing system for diagnostic purposes. The device for the Roche sequencing instrument is expected to launch in 2016 for clinical research, and later as an in vitro diagnostics instrument. The Sequel system features redesigned SMRT cells, which each contains 1 million zero-mode waveguides, as compared to the 150,000 ZMWs in the RS II device. The increased number of ZMWs per SMRT cell enables an approximately seven-fold improvement in throughput, so the total will range between 5 gb and 10 gb and the system has an average read lengths between 8 kb and 12 kb. Like the RS II, the Sequel can run between 1 and 16 SMRT cells per run. Sample preparation will be the same on both instruments, taking around six hours to prepare a library. The chemistry is the same as in the RS II, therefore the consensus accuracy will be the same. The sequel is equipped with improved optics, with computer system that processes more data. It supports all applications that were used in the RS II device, including transcript and base modification analysis.

Although the consumables cost per SMRT cell will increase on the Sequel (to around $700 from about $250 on the RS II), the scientists will obtain a seven-fold improvement in overall throughput. Assuming 7 gb of data per SMRT cell, PacBio estimates that sequencing a human genome at 10 coverage-which should be deep enough to do structural variant analysis-would cost around $3,000. A 30x human genome would cost around $10,000, and a 50x human genome would cost $15,000. (For comparison, in the previous RS II system the SMRT cell contains 150.000 ZMWs, with the throughput of 1 gb per SMRT cell, a 30x human genome would cost around $22,500, while a 50x genome would cost $37,500).

The price to sequence a human genome on the sequel will be still higher, compared to sequencing on Illumina, but the cost for computer and analysis time may be somewhat lower due to longer reads and resulting easier assembly. With its long reads, it will be used in projects for rapid and cost-effective generation of whole genome de novo assemblies, particularly of plant genomes, or in those projects analyzing hard-to-sequence regions of the human genome. The diagnostic instrument (in operation in 2016, with Roche collaboration) is planned for the clinical market.

Schema of operation, the numerous applications in analysis of genes and genomes, relevant publications, collaborations and operation protocols involving the PacBio system, are listed at the web site of the company (www.pacbio.com). Opinion of users and latest news can be found on the Genome Web site (www.genomeweb.com).

Complete Genomics and BGI

Complete Genomics, founded in 2006, plans to become one of the leaders in whole human genome sequencing. Its technology [3] is based upon shearing the Human DNA into pieces about 400 bases long, then using rolling circle amplification to produce so called Nanoballs, which are then spotted upon a patterned substrate. The patterned substrate with nanoballs is subsequently exposed to a repeated proprietary hybridization technique by ligation with fluorescent labeled probes. From the fluorescence signals resulting from the repeated hybridization on the nanoballs the base sequence is determined. The company has developed for its technique the sequencing instruments, chemistry, and software. Complete genomics has sequenced more than 20,000 whole human genomes over the past five years. In a study published in Nature [4], Complete Genomics demonstrated its long fragment read technology, which enables whole genome sequencing and haplotyping from 10 to 20 cells with an error rate of one in 10 million bases, which it hopes would enable truly clinical grade genomes.

In March 2013 Complete Genomics was acquired by BGI-Shenzhen, the world's largest genomics services company, with headquarter in Shenzhen, China. It provides comprehensive sequencing and bioinformatics services for commercial science, medical, agricultural and environmental applications, and collaborates globally in many projects with leading research institutions. Prior to the acquisition by BGI, the Complete Genomics Company was focused on selling whole genome sequencing services to over 150 research customers. It focuses now on building a new generation of high-throughput sequencing technology for clinical and consumer applications. In 2015 Complete Genomics produced and introduced the revelocity system, the end-toend genomics system for large-scale genomes. The revelocity system will use a 300-base insert and mate pair sequencing of 28 bases from each end of the insert. The older technology with the four-adaptor construction had mate pair reads of 35 bases [3]. The remainder of the sequencing technology is the same. DNA nanoballs are formed from the circular pieces of DNA and deposited onto patterned silicon arrays and sequenced by ligation.

Revelocity enables 96 percent of the genome to be covered with an error rate of about 10-6. The system will initially be applied to exome and whole-genome sequencing, but the company plans to develop other applications, like RNA-seq.

In October 2015, BGI has launched a new desktop sequencing platform BGISEQ-500, based on the technology of Complete Genomics. It uses a variation on the "DNA nanoballs" sequencing method, with new engineering features. While Complete Genomics' instruments are applied to sequencing whole human genomes and exomes, the new instrument is more flexible and is planned for RNA sequencing, including in single cells, panels for non-invasive prenatal sequencing, genotyping panels, and other applications. It is a bench top machine with high-throughput, about 200 gigabases per run. The BGISEQ-500 will also be capable of rapid runs producing as little as 8 gigabases of data, an advantage for clinical assays.

BGI and its customers have generated over 250 publications in top journals. Schema of operation, the numerous applications in analysis of genes and genomes, relevant publications, collaborations and operation protocols involving the Complete Genomics and BGI systems, are listed at the web site of the companies (www.completegenomics.com). Opinion of users and latest news can be found on the Genome Web site (www.genomeweb.com).

Oxford Nanopore Platform

The priniciple of the operation of nanopore sequence technique is the analysis of DNA strand directly as the molecule is drawn through a tiny pore suspended in a membrane. Changes in electrical current, or tunneling currents, are used to read off the chain of bases. Great hopes and attention are devoted to the emerging Nanopore technology, because it would allow to sequence Single DNA Molecule in real-time, requiring no amplification step, combined with less expensive hardware and simplicity of operation.

The promise is great, and the progress is finally slowly taking place. When Oxford Nanopore presented the MinIon device in 2012, it was met with slightly pessimistic expectations. In the meantime, many labs worldwide are part of the MinIon analysis and reference consortium (MARC), a group formed by a number of participants in Oxford nanopore's MinIon early-access program (MAP) that plans to conduct a series of projects around the technology. They all wish the technology to be a success, and are assessing how reproducible the device's performance is across laboratories, and developing standard protocols and reference data. The consortium made the raw and aligned nanopore reads of the first test results available recently (corresponding to the state of the MinIon technology in April 2015, www.genomeweb.com, 15.10.2015). The European Nucleotide Archive and the European Bioinformatics Institute coordinated the data distribution and analysis. The goal of the study was to assess the yield, accuracy, and reproducibility of the MinIon by performing replicate experiments at several sites, and to find out which technical factors determine high performance. For the study, the laboratories each generated MinIon sequence data from the same substrain of E. coli using the same protocols.

Overall, using R7.3 flow cells and SQK-MAP005 chemistry, a typical experiment resulted in about 20,000 two-dimensional reads, with a median length of 6.5 kilobases, generating about 115 megabases of data. When they used an 8-kilobase shearing protocol, almost 5 percent of 2D reads were at least 10,000 bases long, some of them more than 50,000 bases. The total error for individual 2D base calls was 12 percent, consisting of 3 percent miscalls, further 4 percent were insertions, and 5 percent deletions. A single run yielded enough 2D bases to cover the E. coli genome 25-fold.

The performance of the MinIon device itself was consistent in the tests, and no experiment failed due to problems with the device. The researchers also did not observe any GC-bias, although they noted this may be hard to detect with an E. coli genome. One reason the quantity and quality of data varied between runs is that many steps in the standard protocol are sensitive to the quality of materials and reagents being used. The results of the MARC report suggest how to improve the performance of the MinIon, including clearer protocol steps, methods for longer library molecules, and improved operation run software scripts.

Overall, MinIon nanopore sequencing has several advantages over short-read sequencing technologies, e.g. its long reads enable sequencing through repetitive regions, its speed generating results in minutes rather than hours, its low capital cost of $1,000 per device, and its small size. Its main limitations are the currently high mismatch and indel error rate and the limited yield per run, ranging from megabases to gigabases. But both will continue to improve, through better bioinformatic tools and new hardware and chemistry.

Oxford Nanopore recently released the new MinIon Mk1 device and reagents, and the company is constructing systems with larger throughput. But the priority is to concentrate on reproducibility and standardization of the performance.

Detection of Structural Variants with Oxford Nanopore Minion Device

Nanopore sequencing can detect structural variants in a mixture of PCR amplicons. In a proof-of-principle the scientists led by Jim Eshleman and Winston Timp from Johns Hopkins University, showed (GenomeWeb.com, 5.10.2015) that they could detect a number of well-characterized structural variants-including large deletions, inversions, and translocations-that affect two tumor suppressor genes in pancreatic cancer cell lines, using amplicon sequence data from Oxford nanopore's MinIon. It shows the ability of nanopore sequencing to correctly and reliably detect SV with only hundreds reads, instead of millions of reads.

In the detection of structural variants, Oxford Nanopore is still behind the other technologies that offer long-range genome information, such as Pacific Biosciences, BioNano Genomics, Illumina synthetic long reads, and OpGen optical mapping. But the MinIon has more potential for diagnostic and screening use. Oxford nanopore platform cost and its operation are very low compared to other systems in use. This will push sequencing to spread from research laboratories to the clinics, where nanopore sequencing could become the tool for the low-level detection of cancer-associated SVs and their early detection. Clinical applications will require improved throughput (compared to amplicon sequencing), needed both for genome-wide detection of SVs and for targeted sequencing of regions likely to contain an SV. Increase in throughput will be accomplished with advances in the current MinIon platform, and with the larger instruments that Oxford Nanopore is developing.

There are now an increasing number of publications using the technique, with applications in analysis of clinical samples and small genomes. Schema of the principle of operation, relevant publications, collaborations and operation protocols involving the Oxford Nanopore system, are listed at the web site of the company (www.oxfordnanopore.com). Opinion of users and latest news can be found on the Genome Web site (www.genomeweb.com).

Other Projects With Protein Nanopores, Solid State and Graphene Nanopores

The nanopore systems in operation (Oxford Nanopore, Genia), use the protein nanopore channels in a membrane. It has been demonstrated that using special protein channels can give resolution of single DNA bases [5-7]. There are efforts underway to develop the nanopores with solid state techniques used in semiconductor industry, which could have some advantages in the mass manufacturing process of chips with large number of nanopores, with the ambition to increase the reproducibility of the nanopore performance. At present the thickness of the solid state membranes is too large, 10-20 nm, to allow the resolution of a single DNA base, since with this thickness there may be 30-50 bases simultaneously inside the nanopore. Further difficulty in the present techniques is the large speed with which the DNA molecule passes through the solid nanopore under the influence of an applied electrical field, making it impossible for the electronic detection system to achieve the single base resolution.

To slow down and control the translocation speed of the single strand DNA through the nanopore, one of the possible way is the use of DNA polymerases, as proposed e.g. in 2010 at EPFL in Lausanne (Ansorge W., note about patent proposal for DNA polymerase as a motor for controlled transport of DNA strand through a nanopore, in an EPFL internal e-mail communication on 4. August 2010 to prof. L. Forro and prof. A. Radenovic, and mailed to prof. Hagan Bayley on 17. August 2010), or in a publication in 2012 [6].

In an effort to reduce the nanopore thickness, several projects were recently described using graphene material for the nanopore [8-11]. The thickness of the graphene nanopore is so small, that just about one third of a nucleotide is inside the pore during the translocation of the DNA molecule through the pore, presenting a theoretical possibility to scan the DNA molecule with a single base resolution. However, as reported in the reports on graphene pores above, there are so far unsolved obstacles. Among them is the large speed of DNA translocation, needing to design much faster electronics signal detection, or alternatively developing some techniques allowing to lower the speed, e.g. by some controlled interactions of the passing DNA strand and the walls of the pore. Reported were also wetting problems of the graphene pore, possibly caused by the hydrophobic surface, requiring perhaps some surface passivation process for the pore.

There were other thin materials tested for the nanopore membranes, e.g. hafnium oxide and boron nitride [11], or molybdenum disulfide [12].

Besides the measurement of the changes in the ionic current as the DNA passes through the pore, there are efforts to measure the electric signals perpendicular to the DNA molecule (called tunneling currents DNA sequencing) as it traverses the pore or channel. The expectations are that each base will show sufficiently different tunneling current, in dependence on its electronic structure, and the base sequence will be deduced from the differences in the tunneling currents.

Genia-Roche

As the various platforms strive toward the $1,000 genome, Genia (www.genia.com) is looking beyond, to the $100 genome. The technology has the potential to reduce the price of sequencing and increase speed, accuracy, and sensitivity by moving away from complex sample preparation and optical detection. The base of Genia's technology is the biological nanopore, a protein pore embedded in a lipid bilayer membrane. The planar electronic sensor technology enables highly efficient accuracy of current readings, which is a common limitation in nanopore sequencing efforts. Genia's NanoTag sequencing approach, developed in collaboration with Columbia and Harvard University, [13], uses a DNA replication enzyme to sequence a template strand with single base precision, as base-specific engineered tags cleaved by the enzyme are captured by the nanopore. As the cleaved tags travel through the pore, they attenuate the current flow across the membrane in a sequence-dependent manner (Figure 1). Thus it is the tag, not the nucleotide, passing through the pore.

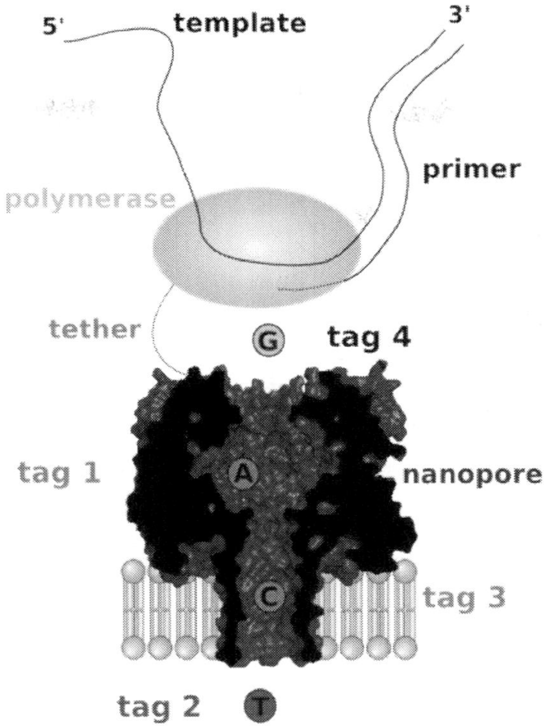

Figure 1: Schematic of a single molecule DNA sequencing by a nanopore with phosphate tagged nucleotides. Genia's NanoTag sequencing approach identifies DNA sequences not by detecting the nucleotides themselves with the nanopore, but by measuring the current changes caused by the passage of each of four di ﬀerent tags that are released from the incorporated nucleotide during the polymerase reaction.

Electrochemically, each of the four tag-types interact with the nanopore recognition site differently, partially blocking the ion current by a characteristic amount which results in a tag-specific electronic signature (Figure 2). DNA sequences are computed from the residual currents measured on the nanopore/DNA complex during the passage of the cleaved tags through the pore. The high sensitivity of electronic detection circuitry underneath each sensor enables to obtain the sequence of single DNA molecules with the Genia platform. The nanotags on the four nucleotides can be in the future optimized with the aim to achieve the highest resolution of the single bases [13].

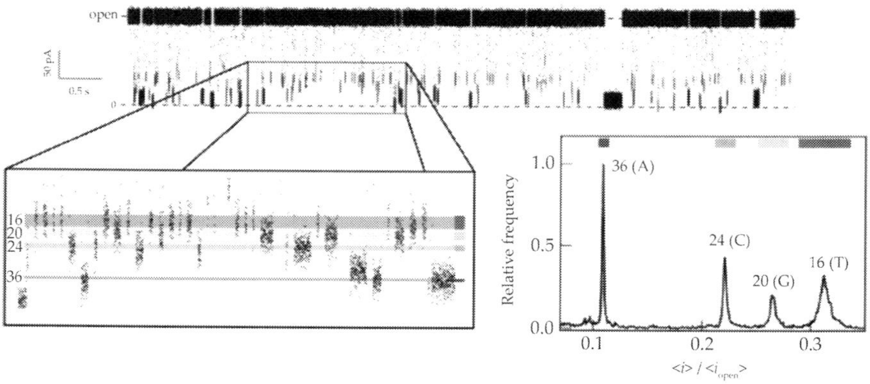

Figure 2: Example using unique current blockade signatures from engineered electronic tags to identify individual nucleotides. From: Scientific Reports (Nature Publications group), "PEG-Labeled Nucleotides and Nanopore Detection for Single Molecule DNA Sequencing by Synthesis" (2,684; DOI:10.1038/srep00684).

Many of today's sequencing platforms rely on specialized, expensive optical sensors. The Genia technology senses changes in electrical currents, detected by standard semiconductors devices, without optical systems. Involving semiconductor technology with the DNA sequencing will reduce the cost of Genia's sequencer operation, making the $100 genome a possibility. The company was acquired recently by Roche.

BioNano Genomics

The BioNano Genomics company is not producing a classical DNA sequencing devices, although it uses rather similar nanochanel technology and fluorescence detection. But by providing a detailed physical Genome Map, it helps to finish sequencing, and to remove sequencing errors caused by repetitive regions. Genome Maps are a complement to sequencing and increase the quality of the results. NGS technologies are essential for nucleotide-level information, but some are limited in long range resolution. The reason is that the fragmented reads are often too short compared to repetitive regions, and longer base reading length would be needed to make a complete map and overcome repetitive regions within and between genes. To detect variation and architecture on a larger scale, scientists need a technology to directly map the genome with a resolution matching the variation in question, e.g. kilobases, not base pairs. Adding a genome map to sequencing data enables to obtain a better view of the whole genome, showing its features in context and functional relationships, across kilobases to megabases. Sequencing can be completed to a higher standard, in less time, with the help of a physical map to facilitate de novo sequence assembly and scaffolding.

The high-resolution Irys System from BioNano Genomics produces a detailed whole-genome map with highly precise anchor-points against which to align data from NGS. Genome maps orient contigs and size gaps by bridging across repeats and other complex elements that break NGS assemblies. The system improves the long-range contiguity across the genome, and helped in many cases to find errors in sequence assembly. Genome Maps represent a very useful validation method for genome contigs and scaffolds in regions assembled from fragment reads.

Structural variations in chromosomal DNA account for a large amount of the variability between human genomes, influencing also phenotypic variations. In the clinical research, structural variants are increasingly associated with several diseases, such as Crohn's disease, autism, schizophrenia, morbid obesity and cancer. To study structural variation, researchers have traditionally been limited to techniques where the genomic DNA is first sheared into very small fragments, which limits the correct assembly of complex genomes. During assembly, repeats and structural relationships between regions are lost by the fragmentation. In the Irys System is the excessive fragmentation avoided, by relaxing and straightening large DNA molecules in nanochanels, so they can be imaged intact. Analysis of large whole molecules and generation of de novo genome maps allows viewing directly the elements in the genome, and to analyse structural variations. Other technologies that offer long-range genome information are Pacific Biosciences, Illumina synthetic long reads, OpGen optical mapping, and recently Oxford nanopore.

In the Irys system, the DNA to be analyzed is labeled at specific sequence motifs for imaging and identification in IrysChips. The labeling method in the system uses a nicking endonuclease to create a single-strand cut in the long DNA molecules, at a specific recognition site, wherever it occurs in the genome. Fluorescently labeled nucleotides are then incorporated by repairing the single-strand nicks with a polymerase. These labeling steps result in a uniquely identifiable sequence-specific pattern of labels to be used for de novo map assembly or for anchoring sequencing contigs. Other labeling techniques may be used in the system. The labeled DNA sample is pipetted onto the IrysChip in the flow cells, movement of DNA in the flowcell is controlled electrophoretically. The DNA stretches in solution, to confine chromosomal-length DNA inside the Nanochannels. The current is transiently turned off and the molecules in solution are stationary and stretched uniformly. At this time imaging is performed, capturing high-resolution, single-molecule images of the labeled DNA, that contains sequence motifs along hundreds of kilobases, (or even over a mega base), in a single contiguous molecule. Once imaged, the molecules are flushed and the process is repeated, allowing imaging of several gigabases of DNA per hour.

This technique reveals meaningful biological information in spite of the presence of repetitive elements, which cannot be resolved by some of the other technologies. It is possible to detect also epigenetic modifications across the genome. In contrast, in techniques where the DNA molecules are sheared to shorter fragments, the biological information may be disrupted or completely lost, e.g. the order and arrangement of functional regions.

Raw image data of labeled long DNA molecules are converted to digital representations of the motif-specific label pattern. These data are then assembled de novo using data analysis software to recreate a whole genome consensus map of the original genome. This genome map enables a variety of analyses, such as sequence finishing and structural variation detection.

Schema of principle of operation, relevant publications, collaborations and operation protocols involving the BioNano Genomics system, are listed at the web site of the company (www.bionanogenomics.com). Opinion of users and latest news can be found on the Genome Web site (www.genomeweb.com).

TECHNIQUES IN DEVELOPMENT

GnuBIO, BioRad

The GnuBIO platform is a new desktop sequencer which incorporates all steps for DNA sequencing into a single system. Utilizing emulsion microfluidics, GnuBIO has developed a scalable DNA sequencing reaction, which encompasses all of the steps necessary for DNA sequencing inside of pico liter-sized aqueous drops. Most other sequencing technologies require separate workflows for target selection, DNA amplification, DNA sequencing and analysis; the GnuBIO technology integrates the complete workflow into a single instrument. The platform provides a single step process which produces genomic result within hours. It is designed to meet the workflow of both research and clinical laboratories, in applications from single genes to whole genomes. GnuBio and its DNA sequencing platform has been acquired by BioRad Company in 2014, (schema of principle operation, and more information see www.gnubio.com).

Qiagen Platform

Qiagen announced several years ago a plan to develop and launch a DNA sequencing platform mainly for clinical applications, to add to its wide range of products for preparation and purification of nucleic acids. For this purpose they bought the DNA sequencing technology from the SBH Company (www.qiagen.com). The complete chain for DNA analysis, from consumables for DNA preparation, over sequencing to data and functional analysis, would present a powerful combination. Although some initial test results were shown, the company has not yet launched the DNA sequencing platform, although all products for the other steps in the analysis chain, including the software for data and function analysis, are already distributed by Qiagen, including its expertise in single cell analysis. Progress, applications and projects can be followed on the web site www.qiagen.com.

Direct Genomics Single-Molecule Sequencing

Direct Genomics, a spin-off of the BGI institute in China, licensed its Single-Molecule Sequencing technique from Caltech, using a sequencing method first developed by Helicos Biosciences, Cambridge, Massachusetts. The GenoCare system will operate with chemistry and reagents from Helicos, but the technique and device were optimized and redesigned. The GenoCare is expected to have a throughput around 10 gigabases per run, run with very high consensus accuracy (approaching 100% by 5x coverage), read length 35 bases initially, and 140 bases with four color chemistry and improvements in hardware. The key improvement compared with the original Helicos technology is the use of total internal reflection fluorescence microscopy (TIRF). The laser is reflected internally and illuminates only very thin region at the flow cell surface with captured DNA, resulting in a significantly improved signal-to-noise ratio. Other improvement is the use of latest generation of CCD camera, with higher sensitivity. The single molecule technique does not require sample preparation module, the DNA capture takes place directly on the flow cell surface. The GenoCare has been designed for clinical applications in China, concentrating on sequencing of gene cancer panels and noninvasive prenatal screening NIPT.

More information on the progress will be available at BGI and Direct Genomics web pages.

POTENTIAL FUTURE TECHNIQUES

DNA Sequencing by Tunneling Currents

Besides the measurement of the changes in the ionic current, as the DNA passes through the pore, there are efforts to measure electric signals perpendicular to the DNA molecule (in a technique called Tunneling Currents DNA Sequencing) as it traverses the pore or channel. The expectations are that each base will show sufficiently different tunneling current and the base sequence will be deduced from the differences in the tunneling currents.

Depending on its electronic structure each base may affect the tunneling current differently, allowing differentiation between the bases during the translocation through the pore, or channel. The technique could have the potential to sequence orders of magnitude faster than the ionic current methods. Sequencing of several DNA oligomers and micro-RNA by the tunneling currents technique has been reported [14,15].

Mechanical Identification and Sequencing of Single DNA Molecule

Recently was presented the proof of concept for a single-molecule platform that allows DNA identification and sequencing [16], based on the mechanical properties of DNA, as explained in the publication. As an entirely novel interesting approach, the technique is not based on the detection of the fluorescent nucleotides, but on DNA hairpin length. By mechanically pulling (in a magnetic tweezers system) on magnetic beads tethered by a DNA hairpin to the surface, the molecule can be un-zipped, and the two DNA strands mechanically separated. In this open state it can hybridize with complementary oligonucleotides, which transiently block the hairpin re-zipping when the pulling force is reduced. By measuring from the surface to the bead of a blocked hairpin, one can determine the position of the hybrid along the molecule, with nearly single-base precision. The approach can be used to identify a DNA fragment of known sequence in a mix of various fragments, and to sequence an unknown DNA fragment by hybridization or ligation. The proof of concept of the technique has been demonstrated, but it may not be easy to develop it into a simple system for routine DNA sequencing applications, competing with other techniques.

Sequencing with Mass Spectrometry

Matrix-assisted laser desorption ionization time-of-flight mass spectrometry, or abbreviated (MALDI-TOF MS), has been investigated specifically as an alternative method to gel electrophoresis for visualizing DNA fragments [17,18]. With this method, DNA fragments generated by chain-termination sequencing reactions are compared by mass rather than by size, and the DNA sequence is deduced. The mass of each nucleotide is different from the others and this difference is detectable by mass spectrometry. Single-nucleotide mutations in a fragment can be more easily detected with MS than by gel electrophoresis alone [17,19].

Early chain-termination and TOF MS methods demonstrated read lengths of up to 100 base pairs. This appeared so far as the limit for the measurable average read size. MS-based DNA sequencing may not be suitable for large de novo sequencing projects, just like chain termination sequencing alone, but it has been used e.g. in bacterial and forensic studies. A recent study did use the short sequence reads and mass spectroscopy to compare single-nucleotide polymorphisms in pathogenic streptococcus strains [20].

DNA Sequencing by Electron Microscopy

This approach directly visualizes the sequence of DNA molecules using electron microscopy. The detection is not straight forward, since the atoms composing the DNA generally are of low mass, thus are not easily visualized by electron microscopic techniques. To be clearly visualized under an electron microscope, the DNA must be labeled with heavy atoms. In addition, specialized imaging techniques and aberration corrected optics are beneficial for obtaining the resolution required to image the labeled DNA molecule. In theory, transmission electron microscopy DNA sequencing could provide extremely long read lengths, but the issue of electron beam damage to the DNA has not been entirely resolved, and the technology has not yet been commercially developed. But the first identification of DNA base pairs within intact DNA molecules has been shown, proving that modified bases containing atoms of increased atomic number were enzymatically incorporated. Direct visualization and identification of individually labeled bases within a synthetic 3,272 base-pair DNA molecule and a 7,249 base-pair viral genome has been demonstrated [21,22].

There are at present two companies working on the DNA sequencing technique by Electron Microscopy, with different approaches. One company, ZS Genetics, published an initial proof-of concept study in 2012 [21], demonstrating that it was possible to label (with a single mercury atom) and identify one of four bases of DNA with an electron microscope. The company claims, it is now able to uniquely label and distinguish between all four bases. The goal for DNA analysis with an EM is that it will enable the long reads and high accuracy. Without the need for fragmentation, the DNA is first labeled, then the EM takes an image of it, while analyzing and distinguishing between the different labels to read out the sequence. If confirmed in the future constructed device, such a strategy could enable reads of around 50,000 base pairs.

ZS genetics uses the transmission electron microscopy (TEM), where the electrons have about a thousand times more energy. In TEM, the high-energy electrons are not scattered enough by the light elements that make up DNA, so the molecule cannot be visualized directly. The heavy metal labeling approach has several problems. One is that the labeling reaction might not go to completion, thus some DNA bases may be missed and not detected. Further problem is that at present it is possible to label only one type of base at a time, because the labels are difficult to distinguish and interfere with each other when they get too close. Thus the sequence is reconstructed from four separate images. In addition, high-energy electrons often damage the DNA, causing the labels to move around, so their position is no longer precisely located on the base.

Another start-up company Electron Optica and its collaborators have experimented and demonstrated in principle, that sequencing of non-labelled DNA by electron microscopy should be feasible. Technical approaches of the two companies differ in several aspects. Electron Optica will use LEEM, where the electrons have low energies up to several hundred electron volts. Because of the lower beam energy in the LEEM device, damage to the DNA molecule will be smaller, and it could reduce sequencing errors. Also, the low-energy electrons are slower, and get scattered by the light elements of DNA (mostly carbon, oxygen, and nitrogen) enabling to obtain sufficient contrast to visualize DNA directly, without the need to label the DNA with heavy metals, as was demonstrated in [22].

They show that the four base types provide sufficient contrast in low-energy electron microscopy (LEEM) for DNA sequencing. Electron Optica uses monochromatic aberration-corrected dual-beam low energy electron microscopy, where two beams illuminate the sample with low-energy electrons, and the reflected electrons compose an image. The LEEM technique has not been used widely in life sciences, because of the somewhat lower resolution, as compared with the other types of electron microscopes. Another reason for the less frequent use of the technique is that most bio-samples are electrical insulators. As such they charge up under the electron beam in the microscope, the charge acquired by the sample affects the trajectory and lowers the resolution of low-energy electrons. To get around this problem, the company had to develop a charge-control technique.

The Electron Optica is designing a new aberration corrector that can bring the resolution to about half a nanometer, needed for a demonstration that all four bases can be distinguished.

On four different kinds of substrates, and involving the use of advanced spectroscopy techniques such as X-ray photo-electron spectroscopy (XPS), and Auger electron spectroscopy, the company achieved contrast needed to distinguish the individual bases. The difference in the contrast among the bases is caused mainly by the differences in the nitrogen content of the bases [22].

The price of a finished sequencing microscope is estimated to be $0.5 million to $1million. The platform would compete with other technologies on read length (the goal is to obtain base readings up to 50 megabases), as well as cost per base, accuracy (error rate of 10-6), and throughput (one genome per day). Besides DNA sequencing, it should also be useful for RNA sequencing and detection of methylation on bases.

More details and progress of the techniques can be found on the web site of the companies, (www.zsgenetics.com, and www.electronoptica.com).

RNAP Sequencing Technique

This method [1,2,23,24] is based on use of RNA polymerase (RNAP), which is attached to a polystyrene bead. One end of the DNA to be sequenced is attached to another bead, and both beads are placed in laser optical traps. RNAP motion during transcription brings the beads closer together, and as their relative distance changes, the change is recorded at a single nucleotide resolution. The sequence is deduced based on the four readouts with lowered concentrations of each of the four nucleotide types, similarly to the Sanger method. Comparison is made between regions and sequence information deduced by comparing the known sequence regions to the unknown sequence regions. Due to the special equipment needed, this technique may be useful in special applications, and may not be easily developed into a wide technique competing with other technologies available.

FRET-based and Raman Spectroscopy Approach for DNA Sequencing

A sequencing-by-synthesis approach that applies fluorescence resonance energy transfer (FRET) has been proposed by VisiGen Biotechnologies (www.visigenbio.com) already more than 10 years ago. In this technique a polymerase containing a donor fluorophore is used in combination with four different acceptor fluorophores for the respective bases. Whenever the polymerase incorporates a nucleotide, a FRET signal is generated through the proximity of the donor and acceptor fluorophores. Sequence information is obtained consequently based on the specific labels for each base. Upon release of the pyrophosphate, which contains the fluorophore, the FRET signal is quenched until the incorporation of the next base. The technique has been very promising, and taken over by Life Technologies in the past, without any news recently concerning its eventual further development.

The use of surface-enhanced raman scattering (SERS), and tipenhanced raman scattering (TERS) techniques for DNA Sequencing are described in a publication [25], as well as in a technical review [26]. A test on a single RNA strand is reported as a promising step towards the development of a novel label-free single-molecule sequencing technique.

PERSPECTIVES FOR FUTURE APPLICATIONS AND DIAGNOSTICS TECHNIQUES

As the techniques and technologies are developing at a fast pace, the DNA sequencing applications are spreading to all areas which are even remotely in connection with DNA analysis. As examples may serve the control of vaccines and medicament purity in pharma industry, where detected are not only active viral components, but also not active dormant viruses. Control of claimed meat quality in food processing industry, forensic analysis, fetal analysis from its DNA circulating in mother`s blood, evolution biology of organisms, are just few several further fields where the technique found application. Here will be mentioned two emerging fields for which the latest DNA sequencing technology is of fundamental importance.

Precision Medicine Personal Genomics

In recent years, precision medicine has gained increasing public attention and support [27]. As a result of technological advances that followed the Human Genome Project, emerged the expectation that the medical field can now include the study of DNA, detecting genetically conditioned illnesses and predicting personalized responses to potential therapies. This new form of medical treatment would allow concentrating on patient’s specific needs in prevention, diagnosis, and treatment. Precision medicine has already revolutionized diagnosis and therapy in a number of cancers. Recognized as a 2013 breakthrough of the year by science magazine, a novel type of cancer immunotherapy has been developed. Genetically modified T-cells taken from blood and re-engineered to detect a protein on cancer cells, are used to fight the specific cancer cells. Compared with chemotherapy, the molecular profile of each individual patient will provide physicians with sufficiently specific details about the genetic condition to enable personalized treatment, minimizing adverse side effects and unnecessary exposure to less efficient treatments. The precision medicine will be also help in devising preventive approaches.

At present a person's genome can be sequenced at a cost of about $1, and some companies already see the price being lowered to around $100. This development indicates that DNA techniques providing individual molecular signatures may replace some traditional tools for diagnosis.

Single Cell DNA Sequencing, Chips for Diagnostics

The isolation and analysis of single cells is a field of increasing interest. The goal of obtaining the sequence of a single DNA molecule from a single cell, ideally the entire length of a chromosome, without amplification and without destruction of the original DNA strand, as discussed in [1,2], has moved closer to reality. The rapid progress in the single cell techniques was described in detail recently in the reviews [28,29]. The technique enables the study of rare cells, and will find applications in microbiology, neurobiology, development, immunology, cancer research, as well as in clinical environment.

Commercial devices for the cell isolation are available (www.fluidigm.com), but most of the analysis still requires amplification steps. After the single cell isolation, recent advances in whole-genome and whole-transcriptome amplification have enabled the sequencing of the low amounts of DNA and RNA present there. The technique made possible studies of genomic and transcriptomic heterogeneity, as they develop in normal as well as disease cells. It will also help to understand the evolution of genomic, epigenomic, and transcriptomic variations as they occur during the life time.

CONCLUSIONS

The diagnostics field has already profited from the new generation of DNA sequencing technology, sample preparation, single-cell isolation techniques, DNA amplification schemes, and computer data analysis with improving software tools. The recent novel therapeutic approaches and public support for efforts in improvements in the healthcare, will no doubt give steady push to develop and innovate further the field. The aims for the technology will be to lower the cost of the equipment and biochemicals involved, increasing simultaneously the reproducibility, reliability and simplicity of the techniques and protocols in operation.

The goal to be achieved, as formulated in [1,2] is still valid, in spite of the great progress in DNA sequencing techniques since then. For genomic sequencing, and for analysis of the ever-more important structural genetic variations in genomes (e.g. copy-number variation CNVs, for chromosomal translocations, inversions, large deletions, insertions, and duplications) it would be a great advantage if sequence read length on the original single DNA molecule could be increased to tens of thousands of bases and more, in a relatively short time. Ideally, the goal would be the sequence determination of a whole chromosome from a single original DNA molecule, from a single cell.

Since the previous reviews, a progress in this field has been very encouraging, particularly in throughput and cost per base on Illumina systems, single-molecule real time approaches on Pacific Bioscience and Oxford Nanopore. Many new developments and improvements are still needed. The promise for the benefits in health care and public interest in the Precision Medicine project, to understand novel therapies, will be the stimuli for continuous efforts in developments of technology, diagnostics and therapies.

REFERENCES

1. Ansorge WJ (2009) Next Generation DNA Sequencing Techniques and Applications. Review in New Journal of Biotechnology N Biotechnoll. 25: 195-203.
2. Ansorge WJ (2010) Novel Next-Generation DNA Sequencing Techniques for Ultra High Throughput Applications in Bio-Medicine, Molecular Diagnostics (2ndedn), ScienceDirect.
3. Drmanac R, Sparks AB, Callow MJ, Halpern AL, Burns NL, et al. (2010) "Human genome sequencing using unchained base reads on self-assembling DNA nanoarrays". Science 327: 78-81.

4. Brock AP, Bahram GK, Andrew BS, Oleg A, Peter H, et al. (2012) Accurate whole-genome sequencing and haplotyping from 10 to 20 human cells. Nature487:190-195.
5. Manrao EA,Derrington IM, Laszlo AH,Langford KW,Hopper MK, et al. (2012) Reading DNA at single-nucleotide resolution with a mutant MspA nanopore and phi29 DNA polymerase, Nature Biotechnol 30: 349-353.
6. Cherf GM, Lieberman KR, Rashid H, Lam CE, Karplus K, et al. (2012) Automated forward and reverse ratcheting of DNA in a nanopore at 5-Å precision. Nature Biotechnol 30: 344-348.
7. Laszlo AH,Derrington IM, Ross BC, Brinkerhoff H, Adey A, et al. (2014) Decoding long nanopore sequencing reads of natural DNA. Nature Biotechnol 32: 829-833.
8. Merchant CA, Healy K, Wanunu M, Ray V, Peterman N, et al. (2010) Nano Lett 10: 2915-2921.
9. Schneider GF (2010) Nano Lett. 10: 3163-3167.
10. Garaj S, Hubbard W, Reina A, Kong J, Branton D, et al. (2010). Nature 467: 190-193.
11. Drndic M (2014) Sequencing with Graphene pores. Nature Nanotechnol 9: 743.
12. Liu K, Feng J, Kis A, Radenovic A (2014) ACS Nano 8: 2504-2511.
13. Kumar S, Tao C, Chien M, Hellner B, Balijepalli A, et al. (2012) PEG-Labeled Nucleotides and Nanopore Detection for Single Molecule DNA Sequencing by Synthesis. Scientific Reports 2, Nature Publications group.
14. Ohshiro T, Matsubara K, Tsutsui M, Furuhashi M, Taniguchi M, et al. (2012) Single-molecule electrical random resequencing of DNA and RNA. Sci Rep 2: 501.
15. Ventra MD (2013) "Fast DNA sequencing by electrical means inches closer". Nanotechnology 24: 342501.
16. Ding F, Manosas M,Spiering MM,Benkovic SJ,Bensimon D, et al. (2012) Single DNA molecule mechanical identification and sequencing. Nature Methods 9: 367-372.
17. Monforte JA, Becker CH (1997) "High-throughput DNA analysis by time-of-flight mass spectrometry". Nature Medicine 3: 360-362.

18. Hall TA, Budowle B, Jiang Y, Blyn L, Eshoo M, et al. (2005) "Base composition analysis of human mitochondrial DNA using electrospray ionization mass spectrometry: A novel tool for the identification and differentiation of humans". Analytical Biochemistry 344: 53-69.
19. Edwards JR, Ruparel H, Ju J (2005) "Mass-spectrometry DNA sequencing". Mutation Research 573: 3-12.
20. Beres SB, Carroll RK, Shea PR, Sitkiewicz I, Martinez-Gutierrez JC, et al. (2010) "Molecular complexity of successive bacterial epidemics deconvoluted by comparative pathogenomics". Proceedings of the National Academy of Sciences 107: 4371-4376.
21. Bell DC, Thomas WK, Murtagh KM, Dionne CA, Graham AC, et al. (2012) "DNA Base Identification by Electron Microscopy". Microscopy and microanalysis, Journal of Microscopy Society of America, Microbeam Analysis Society, Microscopical Society of Canada 18: 1-5.
22. Mankos M, Shadman K, N'diaye AT, Schmid AK, Persson HH, et al. (2012) Progress toward an aberration-corrected low energy electron microscope for DNA sequencing and surface analysis. J Vac Sci Technol B NanotechnolMicroelectron 30.
23. Pareek CS, Smoczynski R, Tretyn A (2011)"Sequencing technologies and genome sequencing". Journal of Applied Genetics 52: 413-435.
24. Fujimori S, Hirai N, Ohashi H, Masuoka K, Nishikimi A, et al. (2012) Next-generation sequencing coupled with a cell-free display technology for high-throughput production of reliable interactome data. Scientific reports 2: 691.
25. Bailo E, Deckert V (2008) Tip-enhanced Raman spectroscopy of single RNA strands: towards a novel direct-sequencing method. Angew Chem Int Ed 47: 1658-1661.
26. Treffer R, Deckert V, (2010) Recent advances in single-molecule sequencing. Current Opinion in Biotechnology 21: 4-11.
27. Collins FS, Varmus H (2015) A New Initiative on Precision Medicine. N Engl J Med 372: 793-795.
28. Wang Y, Navin NE (2015) Advances and Applications of Single Cell Sequencing Technologies. Molecular Cell 58: 598-609.

29. Macaulay IC, Voet T (2014) Single Cell Genomics: Advances and Future Perspectives. PLoS Genet 10: e1004126.

CHAPTER 4

NEXT GENERATION SEQUENCING TECHNOLOGIES: A SHORT REVIEW

Tarek Hamed Attia[1] and Maysaa Abdallah Saeed[2]

[1]Pediatric Department, Faculty of Medicine, Zagazig University, Egypt

[2]Tropical Medicine Department, Faculty of Medicine, Zagazig University, Egypt

ABSTRACT

When the editorial board invited me to write this short review, I remembered my old dream since thirty years ago, is it possible for mankind to live without more health burden? Scientists can answer a lot of difficult questions like this one. The science is growing very fast and next generation sequencing technologies are currently the hottest topic in the field of human and animal genomics researches. In this mini review, I have highlighted some basic knowledge about next generation sequencing technologies and its value for human health.

Keywords

Sequencing technologies, Exome, Genomic captures, Mitochondria

INTRODUCTION

The science is growing very fast and next generation sequencing technologies are currently the hottest topic in the field of human and animal's genomics researches [1]. Early efforts at sequencing genes were painstaking, time consuming, and labor intensive. This situation began to change during the mid-1970s, when researcher Frederick Sanger developed several faster, more efficient techniques to sequence DNA. Indeed, Sanger's work in this area was so ground breaking that it led to his receipt of the Nobel Prize in Chemistry in 1980 [2].

Next-generation sequencing (NGS) is a type of DNA sequencing technology that uses parallel sequencing of multiple small fragments of DNA to determine sequence [3]. In contrast to Sanger sequencing, the speed of sequencing and amounts of DNA sequence data generated with NGS, which is considered a "high-throughput technology", are exponentially greater, and are produced at significantly reduced costs [4]. Third-generation sequencing uses parallel sequencing similar to NGS, but unlike NGS, third-generation sequencing uses single DNA molecules rather than amplified DNA as a template. Thus, third generation sequencing potentially eliminates errors in DNA sequence introduced in the laboratory during the DNA amplification process [5,6].

The starting material that provides a template for clinical NGS is double-stranded nuclear DNA. This can be obtained from a variety of cell types. In some cases, the DNA is further modified in the laboratory to remove non-coding regions when exome sequencing is performed [7,8].

NGS can be used to sequence every nucleotide in an individual's DNA, or limited to smaller portions of the genome such as the exome or a preselected subset of genes [9,10]. Many second generation HTNGS platforms, are commercially available and the race for more additional platforms are continuously on the horizon [1]. The interpretation of the result of clinical NGS laboratories will reported as pathogenic, likely pathogenic, likely benign, benign and of unknown clinical significance [11]. The clinician and patient must be aware that the likelihood of receiving a result of "uncertain significance" is reasonably high, because the clinical implications of many variants are unknown [12].

As defined by the National Society of Genetic Counsellors, genetic counselling is the process of helping people understand and adapt to the medical, psychological and familial implications of genetic contributions to disease. This process helps to understand chance of disease occurrence or recurrence and provide platform for education about inheritance [13]. Consideration of NGS as a clinical tool for genetic diagnosis is appropriate in individuals for whom sequencing of a single gene is unlikely to provide a diagnosis especially in diseases with genetic heterogeneity [14,15]

NGS opens the entire spectrum of genomic alterations for the genetic analysis of complex traits. The main challenge in the fast growing HT-NGS technologies, is to cope with the analysis of vast production of sequencing database through advanced bioinformatics tools [1,16].

AND NOW, WHAT IS THE FUTURE NGS TECHNOLOGIES?

In the near future, HT-NGS technologies with ultra-deep sequencing of genomic DNA will be very helpful for medical purpose like analysis of the disease etiology and development in new drugs [1,17,18]. These researches will be of great value for genetic improvement of animal health and productivity [19].

HT-NGS technologies is a promising tool in the analysis of chromatin immunoprecipitation coupled to DNA microarray (ChIPchip) or sequencing (ChIP-seq), RNA sequencing (RNA-seq), whole genome genotyping, de novo assembling and re-assembling of genome, genome wide structural variation, mutation detection and carrier screening, detection of inherited disorders and complex human diseases, DNA library preparation, paired ends and genomic captures, sequencing of mitochondrial genome and personal genomics [1].

At the end, in the future we have a lot of hopes and a lot of questions.

Do we have the ability to control the pathophysiology of the diseases by understanding in depth their genetic background?

Do we have the ability in the near future to control cancer and prevent its occurrence?

Do we have the capacity in the near future to prolong life expectancy of mankind and control aging?

Finally, I still have a dream, which is an old one, and the answer will come in the near future with the help of sequencing technologies.

REFERENCES

1. Chandra SP, Rafal S, Andrzej T (2011) Sequencing technologies and genome sequencing. J Appl Genet 52: 413-435.
2. Adams J (2008) DNA sequencing technologies. Nature Education 1: 193.
3. Rizzo JM, Buck MJ (2012) Key principles and clinical applications of "next-generation" DNA sequencing. Cancer Prev Res (Phila) 5: 887-900.
4. Voelkerding KV, Dames SA, Durtschi JD (2009) Next-generation sequencing: from basic research to diagnostics. Clin Chem 55: 641-658.
5. Munroe DJ, Harris TJ (2010) Third-generation sequencing fireworks at Marco Island. Nature Biotechnology 28:426-428.
6. Schadt EE, Turner S, Kasarskis A (2010) A window into third-generation sequencing. Hum Mol Genet 19: R227-R240.
7. Mardis ER (2008) The impact of next-generation sequencing technology on genetics. Trends Genet 3: 133-141.
8. Venter JC, Adams MD, Myers EW, Li PW, Mural RJ, et al. (2001) The sequence of the human genome. Science 291: 1304-1351.
9. Mardis ER (2008) Next-generation DNA sequencing methods. Annu Rev Genomics Hum Genet 9: 387-402.
10. Mardis ER (2011) A dacade's perspective on DNA sequencing technology. Nature 470: 198-203.
11. Taylor JC, Martin HC, Lise S,Broxholme J, Cazier JB, et al. (2015) Factors influencing success of clinical genome sequencing across a broad spectrum of disorders. Nat Genet 47: 717-726.
12. Bamshad MJ, Ng SB, Bigham AW,Tabor HK, Emond MJ, et al. (2011) Exome sequencing as a tool for Mendelian disease gene discovery. Nat Rev Genet 12: 745-755.
13. National Society of Genetic Counselors Website.
14. Lee H, Deignan JL, Dorrani N,Strom SP, Kantarci S, et al. (2014) Clinical exome sequencing for genetic identification of rare Mendelian disorders. JAMA 312: 1880-1887.
15. Gilissen C, Hehir-Kwa JY, Thung DT, van de Vorst M, van Bon BW,

et al. (2014) Genome sequencing identifies major causes of severe intellectual disability. Nature 511: 344-347.

16. Day-Williams AG, Zeggini E (2010)The effect of next-generation sequencing technology on complex trait research. Eur J Clin Invest 41: 561-567.
17. Meyerson M, Gabriel S, Getz G (2010) Advances in understanding cancer genomes through second-generation sequencing. Nat Rev Genet 11: 685-696.
18. Bowne SJ, Sullivan LS, Koboldt DC, Ding L,Fulton R, et al. (2011) Identification of disease-causing mutations in autosomal dominant retinitis pigmentosa (adRP) using next-generation DNA sequencing. Invest Ophthalmol Vis Sci 52: 494-503.
19. Green RD, Qureshi MA, Long JA, Burfening PJ, Hamernik DL (2007) Identifying the Future Needs for Long-Term USDA Efforts in Agricultural Animal Genomics. Int J Biol Sci 3: 185-191.

CHAPTER 5

TRANSCRIPTOME ANALYSIS ON RNA-SEQ DATA

Chandra S Pareek[1,2]

[1]Functional Genomics Lab. Faculty of Biology and Environmental Protection, Nicolaus Copernicus University, Torun, Poland

[2]Interdisciplinary Centre of Modern Technology, Nicolaus Copernicus University, Torun, Poland

ABSTRACT

Global researches in the field of biology, biotechnology and medicine require utilization of advanced transcriptome analysis to investigate cellular state, physiology and other relevant biological events. In past decade, RNA-seq is emerged as one of the most versatile application of Next-generation Genome Sequencing (NGS) technology and revolutionized the global researches on transcriptome. The high-throughput RNA-seq data continues to provide unparalleled insight into transcriptome complexity. Now one can consider that the "gold standard" for assessing global transcriptome analysis on RNAseq data is poised to revolutionize our understanding of transcription and posttranscriptional regulation of RNA.

Global researches in the field of biology, biotechnology and medicine require utilization of advanced transcriptome analysis to investigate cellular state, physiology and other relevant biological events. In past decade, RNA-seq is emerged as one of the most versatile application of Next-generation Genome Sequencing (NGS) technology and revolutionized the global researches on transcriptome [1]. The high-throughput RNA-seq data continues to provide unparalleled insight into transcriptome complexity [2]. Now one can consider that the "gold standard" for assessing global transcriptome analysis on RNAseq data is poised to revolutionize our understanding of transcription and posttranscriptional regulation of RNA [3].

Since its first introduction in 2005, tremendous progress has been made both in advancement of NGS platforms (Illumina, Roche, life technologies etc.), as well as in advancement of bioinformatics tools for RNA-seq data analysis. Among the sequencing platforms, Illumina led the way by offering the most versatile equipment, viz., MiSeq, HiSeq series and NextSeq 500/550 [4] for comprehensive transcriptome analysis in mammals. Principally for most of the NGS platforms, the laboratory procedure to generate RNA-seq data is the same involving three major steps: i) isolation of total RNA and/or mRNA ii) library preparations iii) NGS sequencing and production of high-throughput raw sequencing data of transcriptome [5].

However, success of any RNA-seq experiment is critically and solely depends upon the effectiveness of utilized transcriptome analysis methods [6]. That is why with the advancement of novel technologies, the comprehensive nature of generated RNA-seq data has been a boon in terms of transcript identification but the transcriptome analysis is still a big challenge. In general, transcriptome analysis on RNA-seq data were enabled to perform following key tasks:

i. detection of rare and novel transcripts (SNPs and Indels) [7],

ii. determination of the structure of genes (gene counts),

iii. quantification of the changing expression levels of individual transcript and its comparison (DEG) [8],

iv. identification of splicing variants and other post transcriptional modifications [3], and

v. transcriptome mapping or transcript assembly, (de novo assembling and re-sequencing) [9] and

vi. transcriptome annotation [10].

To date, an updated review of literature records PubMed Central (PMC) on transcriptome analysis using RNA-seq technology revealed a total 17221 publications. Majority of publications were on human transcriptome analysis (14760). By applying the major applications of RNA-seq as the keywords, publications can be further categorized as: mapping with 9641, assembly with 7651, de novo assembly with 4981, splicing variants with 5353, Differentially Expressed Gene (DEG) with 370, rare and novel variant analysis (SNPs and Indels) with 3161, and transcript annotation with 8258 publications, respectively.

In general, transcriptome analysis on RNA-seq data starts with raw sequencing data generated from NGS platforms. A single run of any sequencing platform yields an appreciable amount (hundreds of gigabytes) of sequencing data. Before starting into transcriptome analysis on RNA-seq data, one should make sure that the necessary computing, data storage resources and basic bioinformatics expertise are already in place and well set. Generally, RNA-seq raw data come with errors and should be preprocessed before being utilized into downstream analyses such as mapping, assembling etc. At the beginning, the basic tasks such as adapter removal, duplicate quantification and summary statistics on quality score can be performed by standard tools like the fastQC toolkit [11]

Another crucial component for transcriptome analysis of RNAseq data is the normalization [12,13]. When comparing transcripts of different length, it is important to control for length, as a longer transcript will be covered with more reads than an equally expressed shorter transcript. Therefore, a simple way to handle this issue is to divide read counts by the total number of mapped reads (or quantiles of mapped reads). This basic normalization controlling for transcript length and sequencing effort is captured by the commonly used measures such as RPKM (reads per kilo-base exon per million reads) or FPKM (fragments per kilo-base exon per million reads) [14]. Apart from normalization, it is also important to find a statistical distribution approximating the nature of the RNA-seq data. For this, several software packages such as DE-seq [15], bay-seq [16], edgeR [17], NOI-seq [18], are widely utilized. A successful transcriptome analysis on RNA-seq data can also yield a set of candidate genes (CGs), SNPs and ESTs databases that differ between treatments or populations. This type of transcriptome analysis particularly has great significance in experiments involving active breeding populations of plants and animals. For instance, most of the farm animals and plants, CGs, SNPs and ESTs of economic traits and the trait-associated QTLs (Animal qtldb) [19], SNPs and ESTs database are well set and updated.

Apart from its great potential and versatility, transcriptome analysis on RNA-seq data still has some drawbacks for instances, i) the existence of artifacts and biases (regional GC content, preferential sites of fragmentation, and read "pile-up" due to primer affinity and transcript end effects) need to be identified and controlled for, ii) although many methods have been developed to effectively analyze the RNA-Seq data, but improvements are needed to deal with problems associated with multi-reads and estimating the abundance of splice variants.

In a recent developments on transcriptome analysis on RNAseq, research communities are designing, developing and offering the users' friendly and open-access cloud computing web resources [20] that can cover tool installation, relevant file formats, reference genomes, mapping, assembly, transcriptome annotations, qualitycontrol strategies, expression, differential expression, and alternative splicing analysis methods. Another multi-task cloud computing can also be performed by implementing a complete but modular analysis workflow pipeline termed as RAP (RNA-Seq Analysis Pipeline). Using RAP cloud computing, the user can perform a complete transcriptome analysis without any specific technical competence as well as by directly managing the complexity of distributed computational resources. Moreover RAP cloud computing offers a web interface for results management and visualization, by allowing the user to browse and filter the massive amount of data obtained from typical RNA-Seq experiments [21]. With the availability of more and more increasing volume of transcriptome analysis data from large-scale RNA-Seq studies [22], developed "Stormbow" a cloud-based software package, to process large volumes of RNA-Seq data in parallel. Similarly, the user-Friendly gene eXpression analytic tool (FX), which also runs in parallel on cloud computing infrastructure, was developed for the cloud computing of gene expression levels and genomic variant calling [23]. Finally, one can conclude that global transcriptome analysis on RNA-seq data continuously evolve and within the next few years, it will without doubt be exploited to a larger extent and lead to many more innovating discoveries to understand the complexity of mammalian transcriptome.

REFERENCES

1. Pareek CS, Smoczynski R, Tretyn A (2011) Sequencing technologies and genome sequencing. J Appl Genet. 52:413-435.
2. Calabrese C, Mangiulli M, Manzari C, Paluscio AM, Caratozzolo MF, et al. (2013) A platform independent RNA-Seq protocol for the detection of transcriptome complexity. BMC Genomics 14: 855.
3. Spangenberg L, Shigunov P, Abud AP, Cofré AR, Stimamiglio MA, et al. (2013) Polysome profiling shows extensive posttranscriptional regulation during human adipocyte stem cell differentiation into adipocytes. Stem Cell Res 11:902-912.
4. http://www.illumina.com/systems/sequencing-platform-comparison.html
5. Pareek CS (2014) An overview of next generation genome sequencing platforms. In Next Generation Sequencing: Current Technologies and Applications. Caister Academic Press, Canada.
6. Hoeijmakers WA, Bártfai R, Stunnenberg HG (2013) Transcriptome analysis using RNA-Seq. Methods Mol Biol 923:221-239.
7. Quinn EM, Cormican P, Kenny EM, Hill M, Anney R, et al. (2013) Development of strategies for SNP detection in RNA-seq data: application to lymphoblastoid cell lines and evaluation using 1000 Genomes data. PLoS One. 8:e58815.
8. Rajkumar AP, Qvist P, Lazarus R, Lescai F, Ju J, et al. (2015) Experimental validation of methods for differential gene expression analysis and sample pooling in RNA-seq. BMC Genomics 16:548.

9. Martin JA, Wang Z (2011) Next-generation transcriptome assembly. Nat Rev Genet 12:671-682.

10. Trapnell C, Williams BA, Pertea G, Mortazavi A, Kwan G, et al. (2010) Transcript assembly and quantification by RNA-Seq reveals unannotated transcripts and isoform switching during cell differentiation. Nat Biotechnol 28:511-515.

11. Kroll KW, Mokaram NE, Pelletier AR, Frankhouser DE, Westphal MS, et al. (2014) Quality Control for RNA-Seq (QuaCRS): An Integrated Quality Control Pipeline. Cancer Inform 13:7-14.

12. Zhou Y, Lin N, Zhang B (2014)An iteration normalization and test method for differential expression analysis of RNA-seq data. BioData Min 7: 15.

13. Filloux C, Cédric M, Romain P, Lionel F, Christophe K, et al. (2014) An integrative method to normalize RNA-Seq data. BMC Bioinformatics 15:188.

14. Hart T, Komori HK, LaMere S, Podshivalova K, Salomon DR (2013) Finding the active genes in deep RNA-seq gene expression studies. BMC Genomics 14:778.

15. Tang M, Sun J, Shimizu K, Kadota K (2015) Evaluation of methods for differential expression analysis on multi-group RNA-seq count data. BMC Bioinformatics 16:361.

16. Hardcastle TJ (2012) Empirical Bayesian analysis of patterns of differential expression in count data. R package version 2.4.1.

17. Robinson MD, McCarthy DJ, Smyth GK (2010) edgeR: a Bioconductor package for differential expression analysis of digital gene expression data. Bioinformatics 26:139-140.

18. Khang TF, Lau CY (2015) Getting the most out of RNA-seq data analysis. PeerJ 3:e1360.

19. http://www.animalgenome.org/cgi-bin/QTLdb/index

20. Griffith M, Walker JR, Spies NC, Ainscough BJ, Griffith OL (2015) Informatics for RNA Sequencing: A Web Resource for Analysis on the Cloud. PLoS Comput Biol 11:e1004393.

21. D'Antonio M, D'Onorio De Meo P, Pallocca M, Picardi E, D'Erchia AM, et al. (2015) RAP: RNA-Seq Analysis Pipeline, a new cloud-based NGS web application. BMC Genomics 16:S3.

22. Zhao S, Prenger K, Smith L (2013)Stormbow: A Cloud-Based Tool for Reads Mapping and Expression Quantification in Large-Scale RNA-Seq Studies. ISRN Bioinform 2013:481545.

23. Hong D, Rhie A, Park SS, Lee J, Ju YS, et al. (2012) FX: an RNA-Seq analysis tool on the cloud. Bioinformatics. 28:721-723.

CHAPTER 6

INTERPRETATION OF CLINICAL NEXT-GENERATION SEQUENCING DATA: A HURDLE TO JUMP OVER

Wenhui Laura Li

Department of Pathology and Laboratory Medicine, Children's Hospital Los Angeles, Los Angeles, California, USA

SHORT COMMUNICATION

Next-generation sequencing (NGS) has been revolutionary for the clinical diagnostics field. With its high throughput sequencing power and plummeting cost, it has been increasingly used in clinical labs. Instead of testing the candidate genes one at a time by Sanger sequencing, now a lab can test a group of candidate genes at the same time using the NGS method. For example, many clinical labs now offer epilepsy gene panel tests that usually sequence 100-500 genes that are known to be causal or have associations with different kinds of epilepsies. Gene panel tests are also offered for genetic heterogeneous diseases like neurodevelopmental disease, cardiomyopathy, immunodeficiency disease, etc. This approach dramatically increases diagnostic efficiency and helps clinicians to zoom in on the genetic cause for a certain disease in a timely manner. In addition, the NGS technology is also used to diagnose patients who have been through diagnostic odysseys. Exome sequencing is currently used for this purpose in clinical labs. Exome test sequences an exome that contains all the protein coding regions which comprises 1.5% of the genome but contains 80% of recognized disease-causing mutations. Numerous examples illustrated that the exome sequencing method can efficiently identify genetic causes for undiagnosed diseases, which not only helps clinicians to obtain accurate diagnoses but also guides clinicians in the personalized care and treatment of their patients [1,2].

Although NGS technology transforms the clinical diagnostics field, it is still not easy for routine clinical molecular labs to adopt this technology. The hurdle is the interpretation of clinical NGS data, which usually has gigabytes or terabytes of data. Normally, a clinical exome sequencing test detects 20,000-30,000 variants in protein coding regions per patient, and identifying a disease-causing variant from this large number of variants poses a serious challenge for routine clinical molecular labs which usually only deal with one gene and several variants at a time. Without a bioinformatics team, which helps with the sequence alignment, variant calling, and variant filtration, it is impossible for the lab directors to make sense of the huge amount of NGS data, let alone to make clinical interpretations out of it. Commercial software is available; however, often there is an annual subscription fee and a fee for analysis per sample which a small clinical lab cannot afford.

In addition, even after bioinformatics data analysis and filtration, NGS data still needs manual interpretation of those identified genes and variants. Big academic labs such as Baylor Genetics Lab and UCLA Clinical Exome Sequencing Lab have an NGS data review board which includes lab directors, physicians, and researchers in the molecular biology and genetics fields to manually review exome sequencing data and make a final clinical interpretation. This manual review process makes sure that the genetic variant found can explain the clinical presentations of the patient and inheritance pattern of the disease if it is known. Recently bioinformatics pipelines were reported to be able to efficiently prioritize the genetic variants according to the phenotype information and inheritance pattern of disease causal genes [3-5].

The clinical labs that already offer clinical exome sequencing tests have shown the diagnostic yield of the clinical exome is just around 25% [6-9]. The remaining 75% of clinical cases still could not be diagnosed, preventing appropriate treatment of these patients. The explanation of this low diagnostic yield could be the following:

- Disease-causing variants are located outside of protein-coding regions, such as gene promoter, enhancer, deep intronic region, noncoding intergenic region, etc [10]. Not knowing much about these non-protein-coding regions prevents the interpretation of the functional impact of variations seen in them. For example, a gene promoter is important for gene expression, and it is well known that gene promoter mutations can lead to genetic diseases or cancer [11-14]. However, the pathogenicity of a promoter variant could not be easily inferred without doing functional analysis.

- Disease-causing genes or variants are novel or with few functional studies and associated clinical reports; therefore, it is difficult for clinical labs to perform clinical interpretations on those genes or variants. Often, exome-sequencing-identified variants that were not seen previously, the majority of which are missense mutations, and the pathogenicity of variants remains to be tested functionally. In addition, some novel genes are found in exomes that might be associated with human diseases (inferred from animal studies); however, they are not documented in the OMIM and HGMD databases so that clinical interpretation on the variants seen in these genes is impossible.

- Multigenic cause contributes to the disease onset. Currently the clinical labs only analyze genomic data according to the simple Mendelian inheritance pattern. However, the disease etiology could be multigenic.

- Genetic cause is not enough for the disease onset; environmental insult also plays an important role in disease manifestation. Up to date, how the genetic and environment factors interact with each other to initiate disease onset is not clear yet; however, active research is being done for diseases like autism, allergy, and cancer [15-18]. The knowledge gleaned from this research undoubtedly helps us understand these complicated diseases.

In conclusion, with increasing adoption of genomic sequencing technology in clinical labs, the main challenge is to interpret the clinical genomic data accurately in a timely manner so that the lab can give back helpful clinical reports to physicians. The main obstacle for the interpretation of genomic data is due to the unreadiness of processing huge amounts of data, lack of knowledge of genetic variation in normal populations, insufficient clinical and research studies on important disease genes and variants, and unknown functionalities of non-protein-coding regions in the genome.

Currently, genomic data processing is getting more efficient due to a lot of research effort has been put in, and thus more advanced software tools have been made available for free usage. Such software tools can be adapted for clinical usage with the mind of patient privacy, and automatic clinical report upload to patients' medical record.

In addition, for clinical testing, each patient's clinical features have to be considered for data analysis too. With more and more applications of whole genome sequencing, transcriptome sequencing, and epigenomic sequencing, cutting-edge bioinformatics tools are also being created to integrate various genomic data so that clinical lab directors can process complete genomic profile of patients before final clinical interpretations.

In addition, big population sequencing projects, sequencing projects on specific patient cohorts and functional studies on non-coding regions of the genome will allow us, in the near future, to discover the meaning of genomic codes that we do not yet understand, and unveil the secrets hidden in our genome. With the advancement of our tools and knowledge in genomics, we will accumulate unprecedented capabilities to interpret our genome, and thus unleash the full potential of NGS technology.

REFERENCES

1. Worthey EA, Mayer AN, Syverson GD, Helbling D, Bonacci BB, et al. (2011) Making a definitive diagnosis: successful clinical application of whole exome sequencing in a child with intractable inflammatory bowel disease. Genet Med 13: 255-262.
2. Rabbani B, Tekin M, Mahdieh N (2014) The promise of whole-exome sequencing in medical genetics. J Hum Genet59: 5-15.
3. Smedley D, Jacobsen JO, Jager M, Kohler S, Holtgrewe M, et al. (2015) Next-generation diagnostics and disease-gene discovery

with the Exomiser. Natprotoc10: 2004-2015.

4. Smedley D, Robinson PN (2015) Phenotype-driven strategies for exome prioritization of human Mendelian disease genes. Genome Med7: 81.
5. Yang H, Robinson PN, Wang K (2015) Phenolyzer: phenotype-based prioritization of candidate genes for human diseases. Nat Methods12: 841-843.
6. Lee H, Deignan JL, Dorrani N, Strom SP, Kantarci S, et al. (2014) Clinical exome sequencing for genetic identification of rare Mendelian disorders. JAMA312: 1880-1887.
7. Yang Y, Muzny DM, Reid JG, Bainbridge MN, Willis A, et al. (2013) Clinical whole-exome sequencing for the diagnosis of mendelian disorders. N Engl J Med 369: 1502-1511.
8. Gahl WA, Tifft CJ (2011) The NIH Undiagnosed Diseases Program: lessons learned. JAMA 305: 1904-1905.
9. Farwell KD, Shahmirzadi L, El-Khechen D, Powis Z, Chao EC, et al. (2015) Enhanced utility of family-centered diagnostic exome sequencing with inheritance model-based analysis: results from 500 unselected families with undiagnosed genetic conditions. Genet Med 2015 17: 578-586.
10. Diederichs S, Bartsch L, Berkmann JC, Frose K, Heitmann J, et al. (2016) The dark matter of the cancer genome: aberrations in regulatory elements, untranslated regions, splice sites, non-coding RNA and synonymous mutations. EMBO Mol Med8: 442-457.
11. Theuns J, Brouwers N, Engelborghs S, Sleegers K, Bogaerts V, et al. (2006) Promoter mutations that increase amyloid precursor-protein expression are associated with Alzheimer disease. Am J Hum Genet 78: 936-946.
12. Richter S, Vandezande K, Chen N, Zhang K, Sutherland J, et al. (2003) Sensitive and efficient detection of RB1 gene mutations enhances care for families with retinoblastoma. Am J Hum Genet 72: 253-269.
13. Sagi M, Frenkel A, Eilat A, Weinberg N, Frenkel S, et al. (2015) Genetic screening in patients with Retinoblastoma in Israel. Fam Cancer14: 471-480.

14. Cowell JK, Bia B, Akoulitchev A (1996) A novel mutation in the promotor region in a family with a mild form of retinoblastoma indicates the location of a new regulatory domain for the RB1 gene. Oncogene 12: 431-436.
15. Koufaris C, Sismani C (2015) Modulation of the genome and epigenome of individuals susceptible to autism by environmental risk factors. Int J Mol Sci 16: 8699-8718.
16. Vijayakumar NT, Judy MV (2016) Autism spectrum disorders: Integration of the genome, transcriptome and the environment. J Neurol Sci 364: 167-176.
17. Campbell DE, Boyle RJ, Thornton CA, Prescott SL (2015) Mechanisms of allergic disease - environmental and genetic determinants for the development of allergy. Clin Exp Allergy 45: 844-858.
18. Simonds NI, Ghazarian AA, Pimentel CB, Schully SD, Ellison GL, et al. (2016) Review of the Gene-Environment Interaction Literature in Cancer: What Do We Know? Genet Epidemiol 40: 356-365.

CHAPTER 7

NEXT GENERATION SEQUENCING-CURRENT STATUS

Suma J[1] and Sivakumar G[2]

[1]New York Genome Center, New York, USA

[2]Qassim University, College of Applied Medical Sciences, Buraidah, Kingdom of Saudi Arabia

ABSTRACT

Next Generation Sequencing (NGS) is revolutionizing the study of "omics" (e.g., genomics, transcriptomics, methylomics, etc). The sequencing cost has gone down tremendously and thereby allowing most of the methods to deviate from microarrays and traditional Sanger sequencing. Illumina has the highest market share in NGS followed by Thermofisher Scientific, Roche and Pacific Biosciences. They currently sell the following instruments: HiSeqX series, HiSeq series, MiSeq, NextSeq and MiniSeq. With the arrival of HiSeqX, Illumina projects that the cost per Gigabase has gone down from $100,000,000 in the year 2000 to $1-10 in year 2014. Read-length, run time bacterial genome sequencing and maximum throughput on different platforms is reviewed in Van Djik et al., and the number of applications that can be performed using NGS is increasing every day. They can be categorized into three areas as described below.

INTRODUCTION

Next Generation Sequencing (NGS) is revolutionizing the study of "omics" (e.g., genomics, transcriptomics, methylomics, etc). The sequencing cost has gone down tremendously and thereby allowing most of the methods to deviate from microarrays and traditional Sanger sequencing. Illumina has the highest market share in NGS followed by Thermofisher Scientific, Roche and Pacific Biosciences (http://cen.acs.org/articles/92/i33/Next-Gen- Sequencing-Numbers-Game.html). They currently sell the following instruments: HiSeqX series, HiSeq series, MiSeq, NextSeq and MiniSeq. With the arrival of HiSeqX, Illumina projects that the cost per Gigabase has gone down from $100,000,000 in the year 2000 to $1-10 in year 2014 [1]. Read-length, run time bacterial genome sequencing and maximum throughput on different platforms is reviewed in Van Djik et al., and the number of applications that can be performed using NGS is increasing every day. They can be categorized into three areas as described below [2].

NGS APPLICATIONS

Genomic Sequencing

With the drastic drop in the cost of sequencing, we can sequence the whole genomes since the human genome project. Whole Genome Sequencing (WGS) is used to characterize genetic markers associated with diseases. WGS has revolutionized our understanding of the relationship between genomic variants and its phenotypes. WGS in combination with long read sequencing is used to assemble novel genomes referring to as DeNovo sequencing. With the advance in technologies, WGS can now be done at a single cell level. Few among many single cell genomics applications include reconstructing cell lineage and understanding tumor heterogeneity. Since many genomic variants that are associated with diseases occur in exome, researchers minimize sequencing cost by targeted sequencing: enriching for exons (Whole exome sequencing) or by multiplex PCR.

RNA SEQUENCING

Similar to WGS, mRNA sequencing can also be done on cells in bulk as well as at single cell level. WGS when coupled with mRNA sequencing allows us to understand the effect of a variant on gene expression. Total RNA-Seq after depletion of ribosomal RNA will not only shed light on exonic expression levels but also on intronic levels.

LOCATION-BASED TECHNIQUES

To better understand the transcription and translation processes, location based techniques such as ChIP-Seq (chromatin immunoprecipitation sequencing) and CLIP-Seq (crosslinking immunoprecipitation) have been developed to identify protein-DNA interactions and protein-RNA interactions respectively. Other methods to identify RNA-protein interactions include iCLIP and PAR-CLIP. Chromosome conformation capture (3C)-based systems such as 4C, 5C,6C, Hi-C and ChIA-PET has also been developed to identify DNADNA interactions. These methods help in 3D-construction of the genome. Also, methods for epigenetic measurements such as DNaseISeq, ATAC-Seq and bisulfite sequencing have been established. While DNaseI-Seq and ATAC-Seq provide information on the openness and structure of chromatin; bisulfate sequencing shed light on DNA methylation. Recently, ATAC-Seq and bisulfate sequencing were developed for single-cell analysis [3,4].

LONG MOLECULE SEQUENCING

While Illumina is great for sequencing short inserts, long inserts are needed for repetitive regions in genomes which cannot be easily amplified. Key players in long range sequencing are Pacific Biosciences, 10X Genomics, and Oxford Nanopore Technologies. Sample and library preparation cost.

Even though the cost for sequencing is coming down, the challenge remains to bring down the cost of sample and library preparation. Especially for highly multiplexed and low coverage sequencing purposes, the cost per sample of library preparation overweighs the cost per sample of sequencing. To address this challenge, researchers are coming up with methods that can be performed in 96 or 384-well plate formats. Automating sample and library preparation will save a lot of time and thereby bring down the effective cost per preparation. Traditional library preparation includes shearing, end-repair, A-tailing, ligating and amplification. It takes around 1-2 days to perform a library preparation. An exciting alternative to this traditional library preparation method is Transposase based library construction that can be completed in few hours. Illumina commercializes this kit as Nextera. For low coverage sequencing purposes, several multiplexing strategies have very well addressed this challenge. Recent publications demonstrate split-pool barcoding approach for multiplexing thousands of single cells [5,6].

NGS IN CLINICS

With the advent of many technologies in NGS and a drastic drop in the cost of sequencing, NGS has now made its way into clinics. It is changing the way the diseases are diagnosed. One of the most common applications is the cancer diagnosis. Until recently, cancers are diagnosed based on morphological criteria, but a huge effort is undertaken to shift the paradigm of diagnosis based on genetic mutations and fusions [7]. Another well-known application is NIPT (Non-invasive prenatal testing). Many companies are now offering tests such as Natera, Verifi, MaterniT21 for NGS-based NIPT. NGS has a role in personalized medicine. Personalized medicine has a control over individualizing medicine both in terms of treatment and diagnosis. NGS also has the potential to accelerate the early detection of disorders and toxicity and to detect pharmacogenetics markers to customize treatments [8-10].

COMPUTATIONAL CHALLENGES

With most of the applications tending towards NGS, excess data is being generated. As a result of this, the cost of storing data is considerably increasing than that of the cost of sequencing to produce equivalent amount of data. Also, it is becoming a real challenge to handle the huge amount of data generated in an efficient manner both regarding storage and analysis. One of the possible solutions that are being explored is the usage of cloud-based storage. One such platform that Illumina provides is BaseSpace and can be used for both storages as well as analysis. For now, BaseSpace with Amazon Web Services cloud computing is making an effort to minimize the problem of NGS data storage, processing and interpretation [11].

REFERENCES

1. An Introduction to Next-Generation Sequencing Technology Table of Contents (2015) 1-16.
2. Van Dijk EL, Auger H, Jaszczyszyn Y, Thermes C (2014) Ten years of next-generation sequencing technology. Trends Genet 30: 1-9.
3. Buenrostro JD, Wu B, Litzenburger UM, Ruff D, Gonzales ML, et al. (2015) Single-cell chromatin accessibility reveals principles of regulatory variation. Nature 523: 486-490.
4. Smallwood SA, Lee HJ, Angermueller C, Krueger F, Saadeh H, et al. (2014) Single-cell genome-wide bisulfite sequencing for assessing epigenetic heterogeneity. Nat Methods 11: 817-820.
5. Klein AM, Mazutis L, Akartuna I, Tallapragada N, Veres A, et al. (2015) Droplet Barcoding for Single-Cell Transcriptomics Applied to Embryonic Stem Cells. Cell 161: 1187-1201.
6. Macosko EZ, Basu A, Satija R, Nemesh J, Shekhar K, et al. (2015) Highly Parallel Genome-wide Expression Profiling of Individual Cells Using Nanoliter Droplets. Cell 161: 1202-1214.
7. Gagan J, Van Allen EM (2015) Next-generation sequencing to guide cancer therapy. Genome Med 7: 80.
8. Gonzalez-Garay ML (2014) The road from next-generation sequencing to personalized medicine. Per Med 11: 523-544.
9. Gowder SJT (2013) An updated review of toxicity of bisphenol A (BPA) with special reference to the kidney. Current Mol Pharm 6: 163-172.
10. Greek R (2013) Animal Models in Drug Development, New Insights into Toxicity and Drug Testing, InTech, Croatia.
11. Wandelt S, Rheinländer A, Bux M, Thalheim L, Haldemann B, et al. (2012) Data Management Challenges in Next Generation Sequencing. Datenbank-Spektrum 12: 161-171.

CHAPTER 8

THE PHARMACEUTICAL APPLICATIONS OF NEXT GENERATION SEQUENCING IN ONCOLOGY DRUG DESIGNING AND DEVELOPMENT

Dhuha H Nawab

Pharmacy College, Umm Al Qura University, Saudi Arabia

ABSTRACT

Intensive efforts and international collaborations have been performed and are still ongoing in the fight to beat cancer, and many successes have been achieved as a result of hardworking scientists. Although cancer still exists, new information is learned every day, and research continues to be published containing new theories and findings. Cancer research is characterized by the use of advanced technologies that play a role in accelerating results, which then translates into clinical practice to improve the concept of personalized medicine. Currently, next-generation sequencing [NGS] is one of the most advanced technologies used in cancer research due to its ability to provide valuable information from genomic sequencing. Not only is it capable of examining genes but it can also detect and target expected mutations in certain clinical situations and find unexpected sequence variations without consuming substantial time or effort; therefore, NGS provides clinicians with a better understanding of cancer mechanisms. This article demonstrates the pharmaceutical applications of NGS, from tumor markers to pharmacogenomics and targeted therapy, precision medicine, vaccinology, biopharmaceutics, polypharmacology, toxgnostics, and pharmacoepidemiology.

KEYWORDS

Next generation sequencing, NGS, Cancer, Pharmacogenomics, Precision medicine, poly pharmacology, Toxgnostics, Pharmacoepidemiology, Personalized biomarker

INTRODUCTION

Genes, which are present in every cell of the body, are composed of deoxyribonucleic acids, or DNA. DNA is composed of the following four nucleotide bases: adenine, guanine, cytosine and thymine [AGCT sequence]. Contrary to popular belief, the DNA sequences of all individuals are 99.9% identical and only 0.1% unique.

DNA controls vital cell processes in different ways, such as the control of cell growth, division and death, by making specific proteins; therefore, it is important for each gene to have a specific coding "sequence", allowing all of the subsequent proteins to be made properly. Genetic variations commonly occur, but not all genetic variations cause disease.

Genetic variations are simply differences in DNA sequences, and they can be observed at every genetic level in DNA—in the genes, chromosomes, and proteins and in their functions. Therefore, when discussing the variations in a DNA sequence, one is generally referring to mutations that are present at a level of less than 1% in a population.

Generally, two types of genetic mutation events create all forms of variation:

1. Single base mutations [single nucleotide polymorphisms, or SNPs] or
2. The insertion or deletion of one or more nucleotides.

Therefore, if an individual has a mutation (including SNPs, structural variations, copy number variations [CNVs], somatic copy number aberrations [CNAs], insertions, deletions, differentially expressed genes, differentially expressed isoforms, translocations, expressed variations, predicted gene fusions), the presence of these mutations will affect transcription and can alter the protein structure and function, leading to complex diseases, such as **cancer**.

The beginning of all types of cancer, which is a multifactorial disease that is characterized by its intrinsic heterogeneity [1], begins with mutant cells that create abnormal proteins with altered functions or no protein at all, which then begins the carcinogenic process— initiation, promotion, progression and metastasis. In other words, at the normal cellular scale, there is a balance between growth promoters and inhibitors, but in cancer this balance is lost, leading to sequential processes and the phenotypic hallmarks of malignancy. As proliferation increases, difficulties in differentiation lead to the immortalization of the cells and resistance to normal apoptosis.

The human body normally has the ability to correct the majority of these mutations, but in cancer, this ability differs because the mutations begin to accumulate in vital genes, resulting in alternations to major cellular pathways. As more accumulations occur, the cancer worsens.

Some types of cancer can arise from normal tissues or from specialized cells types, but other types do not fit into the major classifications, such as melanomas, which arise from the neural crest, a primitive embryonic structure [2]. It is worth noting that the information provided by a mutant gene will be different from the normal gene. This clearly shows that there is an increased need to use advanced techniques to read DNA sequences and to detect mutations, to better understand these mutations and to determine the best personalized cure based on the results.

Unfortunately, due to the complexity of cancer, there are currently no drugs that can eliminate cancer cells in patients. Many anticancer therapies are taken in combination and at different cycles, and the majority are toxic, with a narrow therapeutic index and the ability to cause morbidity. The first generation of anticancer therapies began in the 1950s; these therapies were mostly cytotoxic [3] and caused several side effects in cancer patients.

Another challenge that lies in cancer treatment is the ability to detect and diagnose cancer during the early stages. The early detection of cancer leads to a significant increase in the cure rate. However, the majority of cancers are detected during their late stages, when treatment opportunities are limited, resulting in poor survival [4]. Therefore, there is a need to utilize more novel approaches for cancer management that allow the targeting of cancer cells without causing harm to normal cells.

The general approaches for cancer management and therapy lie in destroying or removing cancer cells, either via surgery, irradiation, or chemotherapy. The combination of these approaches shows many benefits for improving a cancer patient's quality of life and increasing their survival rate.

Each approach has is important in its own way. Surgery plays multiple roles in cancer management, including prophylaxis if the organ is not crucial for survival; diagnosis, by taking a biopsy from a cancer patient; staging and measuring the extent of the **tumor**, including allowing the determination of the type of tumor, the growth extent, the size, the nodal involvement and the tumor distance and regional spread; as a definitive or curative treatment, by removing the entire tumor and the lymph nodes that are associated with it; as palliative care to reduce the symptoms and distress; as an adjuvant or supportive therapy, such as a tracheostomy and feeding tubes; as a reconstructive or rehabilitative therapy to reduce deformities; and as a salvage treatment for local disease recurrence, such as via a salvage radical cystectomy or mastectomy [5].

Radiation also plays many roles because both cancer cells and normal cells are affected in cancer. The primary function of radiation is as a curative treatment. For example, some cancer types, such as skin cancer, prostate cancer, early stage larynx cancer and Hodgkin's lymphoma, can only be cured with the use of radiation. Additionally, radiation serves as a form of adjuvant treatment after definitive surgery or as a palliative treatment. This approach has undergone substantial developments, such as the use of intensity-modulated radiation therapy [IMRT], image guided radiotherapy [IGRT], radioimmunotherapy, stereotactic radiosurgery [SRS], and stereotactic radiotherapy [SRT] [5].

Chemotherapies have proven to be effective during the early stages of cancer because they interfere with DNA synthesis steps and tumor replication. They play roles as a form of adjuvant treatment to reduce micrometastases and cancer recurrence or as a neoadjuvant to shrink a tumor [5]. An example is the multi-drug treatment protocol used in Hodgkin's lymphoma. ABVD, which consists of doxorubicin, bleomycin, vinblastine, and dacarbazine, works via intercalation, breaking the DNA strand, and microtubule inhibition [2].

NEXT-GENERATION SEQUENCING [NGS]

One of the more recent advances in genomic technologies is the development of high-throughput sequencing platforms. These technologies have become powerful tools in cancer research.

High-throughput sequencing is based on 3 main steps: using a solid support to immobilize the DNA samples, using automated fluidic devices for cyclic sequencing reactions and using imaging to detect molecular events [6].

First-generation sequencing [Sanger sequencing] is considered the gold standard for sequencing, has been used as a powerful tool for genomic analysis in the past, and is still in use for the validation of high-throughput sequencing analyses. High-throughput sequencing encompasses all of the terms used to describe a number of sequencing technologies, including Illumina [Solexa] sequencing, Roche 454 sequencing, and SOLiD and Ion Torrent.

NGS is flexible because it can be combined with other technologies and supplementary tests. TAm-seq, which is a method developed by Forshew et al., has the ability to detect TP53 mutations by combining primer design and NGS [7]. NGS has many advantages over Sanger sequencing. For example, during sample preparation, less DNA is required for the preparation of the sequencing libraries; normally 3 μg of DNA is required to sequence the BRCA1 and BRCA2 genes using Sanger sequencing, but only 500 ng of DNA is required for chipcaptured NGS sequencing [8]. NGS provides the key to understanding the process of oncogenesis in a genetic manner, due to its ability to provide information regarding a genomic sequence. NGS is capable of both detecting expected mutations in a certain clinical situation and determining unexpected sequence variations, therefore playing a significant role in treatment and in other therapeutic purposes [9].

APPLICATIONS OF NGS

Tumor Markers

The annual report from the American Cancer Society provides the number of new cancer cases and deaths each year. In 2014, approximately 1,665,540 new cancer cases were diagnosed, and there were 585,720 cancer deaths. Anyone can develop cancer, but a substantial proportion of cancers can be prevented [10] if diagnosed accurately and early using proper methods and not at the advanced stages of cancer propagation.

What makes cancer so difficult to detect is that the lifespan of most cancers is characterized by silence, which means that all of the changes that occur in normal cells during their conversion into cancer cells, such as dysplasia, hyperplasia and pleomorphism, occur silently [5].

Additionally, a cancer diagnosis can be difficult because the patient may be asymptomatic or have signs and symptoms of other noncancerous diseases; in this case, a diagnostic test is required. Therefore, a biopsy is performed, which requires the surgical removal of a piece of the suspected cancerous region [5].

Currently, a new interest of scientists and researchers is the productions of genome-based blood tests. One of the major benefits of these advanced tests is the detection of tumor markers, which are the materials secreted by a tumor. The development of personalized blood tests for cancer using whole-genome sequencing has been a focus at Johns Hopkins University. Additionally, NeoGenomics, a leading provider of cancer-focused genetic and molecular testing services, announced on Friday, July 11, 2014 that it had launched 23 new types of cancer profiles based on NGS (**Figure 1**).

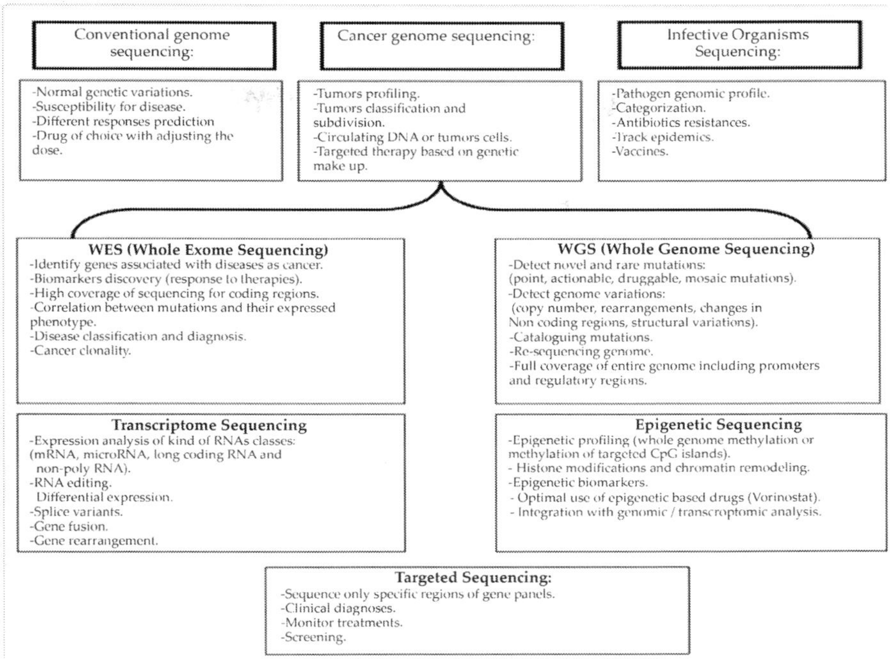

Figure 1: Clinical applications of NGS.

The history of the development of tumor markers began during the 1960s when enzymes, hormones and serum proteins were utilized as tumor markers for therapeutic purposes [11].

During the late 1970s, the first tumor marker was discovered, which contained carcinoembryonic antigen [CEA] or carcinoembryonic protein [12], providing new hope and ambition in the fight against cancer by gaining more valuable information regarding the properties of tumor markers and their mechanisms of action. In addition, this discovery provided insights on how to translate this information into clinical practice, especially in cancer patient management.

Tumor markers can be discovered by detecting SNPs, which are the most commonly occurring mutations in the population. Using the following sequence as an example: ATG CAA to ACG CAA, if a single substitution occurs from T to C, the entire nucleotide sequence will carry the disease because the change in the sequence, or the mutation that occurred, will produce an abnormal protein. Therefore, this SNP is the marker for this disease. The next step is the development of strategies for the discovery and utilization of personalized tumor markers by using individual genetic profiles. Thus far, personalized tumor markers have achieved many successes in different fields of oncology, including early cancer detection, screening, diagnosis, prognosis, targeted therapy, therapeutic response, and monitoring and recurrence (**Table 1**).

Table 1: The critical roles of tumor markers in different cancer types

Cancer type	Tumor marker	Clinical findings
Bladder cancer	ICAM-1/VCAM-1	Higher in all cancer patients than in controls [11].
	sFas ligand	Higher levels of sFasL predict early recurrence in Ta bladder carcinoma [11].
Brain cancer	bFGF	Indicate presence of tumor [11].
	VEGF	Indicate presence of tumor [11].
Cervical cancer	IL-2R	Higher levels in 50% of patients with squamous cell carcinoma [11].
	SCC	Higher levels in 67.5% of patients with squamous cell carcinoma [11].
Colon cancer	Angiogenin	High levels correlated with cancer progression [11].
	E-selectin	Elevated in colon and breast cancer patients [11].
Endometrial cancer	P53	Adjunctive test for gynecological cancers preoperative [11].
	IL-2R	Higher levels in 51.4% of patients with endometrial [11].
Esophagus cancer	IL-2R	Significantly increased in esophagus cancer patients [11].
	P53 AB	Detected in 43% of esophagus cancer patients [11].
Stomach cancer	c-erbB-2	Correlated with HER-2/neu tissue over expression [11].
	COX-2	Over expression is associated with high levels of prostaglandin E2 biosynthesis and angiogenesis [11].

Breast cancer	TGF-β	Highly expressed in primary breast tumors [11].
	ICAM-1	High levels in 96% of sICAM-1 in breast cancer patients [11].
Leukemia	TNF-α	Therapy clinical efficacy, indicator of residual disease presence [11].
	IL-2R	Only elevate it in relapse patients, interferon therapy response prediction [11].
Lung cancer	SVCAM-1 and other CAMs	Detect abnormal levels in SCLC [11].
	TNF-α,TGF-β	Show favorable prognosis for positive lung cancers [11].
Pancreatic cancer	CgA	High levels in 99% carcinoid tumors .elevated in endocrine pancreatic tumor, multiple endocrine neoplasia 1 syndrome [11].
	P53 mutant protein	Detected in 50% of the patients [11].

Pharmacogenomics

Pharmacogenomics is the combination of pharmacology and genomics. This field focuses on the study of how drugs respond differently from person to person due to genomics [13] and how genotype-phenotype information can be utilized in individualized medicine.

The major benefits of PG are as follows:

- The development of drugs to maximize their therapeutic effects.
- The ability to design more accurate dosage methods based on an individual's genetic profile rather than using traditional measurements, such as body weight and age [13].
- The ability to identify the responders /non-responders for different drugs and to determine the risk factors for that drug.

To apply these benefits in real life, one must know the main principles and methodology used in PG, which can be summarized as follows:

- The use of classical genetics techniques [heritability] by applying Mendelian genetics.

- The use of genome-wide association studies [GWAS] to examine genetic differences in different individuals, which is performed by comparing the DNA sequences between two groups—the first group is affected by the disease, and the second group is the control without the disease. If the results indicate the presence of a SNP more frequently in one group than in the other, it is likely associated with the disease.

1. The use of association studies, which consume significant time and effort, especially during follow-up.

These studies connect phenotypes with genetic markers and are based on three primary strategies:

1. Case-control association studies, 2. family-based association studies, and 3. population-based association studies.

Different drugs have different effects in different populations. This indicates that different patients may have the same symptoms, the same findings, and the same disease, and although they are taking the same drug at the same dose, they show different effects and reactions. Therefore, different people react differently to drugs, which are why they are classified as responders, non-responders, or toxic responders due to these individual variations. A percentage of patients benefit from approved drugs, but a portion of drugs have failed in clinical trials, and more importantly, many approved drugs have been removed from the market due to their serious adverse effects.

Some possible reasons for these individual variations may be ethnicity, age, environment, gender and genetic variation. In the field of drug application, the genetic variations that are present in critical genes that are responsible for metabolizing certain drugs and transporting or targeting drugs will influence a drug's activity (**Table 2**).

Table 2: The effect of genetic variations on anticancer drugs and how the FDA uses this information in drug labeling and tests that are approved to detect these mutations

Selected Germline Genetic Variants			
Drug	5-FU	6-MP	Irinotecan
Gene	DPD	TPMT	UGT1A1

Role	Rate limiting metabolicenzyme	Major inactivation enzyme	Major enzyme for SN-38 Glucuronidation.
Mutations	DPYD*2A= partial DPD deficiency lead to 5-FU toxicity [14].	TPMT*2,TPMT*3A,TPMT*3C= Low TMPT activity [3] which may associated with high myelotoxicity risk [15].	UGT1A1*28=SN-38 accumulation leading to Irinotecan toxicity [3].
FDA Label	yes	yes	yes
Labeling Section	Warnings	Dosage and Administration, Contraindications, Precautions, Adverse Reactions, Clinical Pharmacology	Dosage and Administration, Warnings, Clinical Pharmacology
Test	Theraguide 5FU	TPMT Testing	InvaderR UGT1A1
Url	http://www.myriad.com/products/theraguide-5-fu/	http://labtestsonline.org/understanding/analytes/tpmt/tab/test/	http://www.invaderchemistry.com/invader_applications/invader-ugt1a1.html
Selected Somatic Genetic Variants			
Drug	Crizotinib	Vemurafenib	Gefitinib/Erlotinib
Gene	EML4-ALK	BRAF V600E	EGFR-tumor activating
Role	receptor tyrosine kinase that is aberrant in a variety of malignancies	serine-threonine protein kinases which involved in many cellular processes. Mutant BRAF has been implicated in the pathogenesis of several cancers.	receptor tyrosine kinases, induces a conformational change that facilitates receptor homo- or heterodimer formation. Activated EGFR then phosphorylates its substrates, resulting in activation of multiple downstream pathways within the cell.
Mutations	-Resistance in C1156Y,L1196M mutations present [16-17]. -Response in EML4-ALK rearrangement present[18-21].	- Response in BRAF V600E mutation present [22]. -Resistance in MAPK pathway reactivation [23,24].	- Response in EGFR-activating mutation present [25-29]. -Resistance in T790M mutation present [30,31].
FDA Label	yes	yes	yes

	Indications and Usage, Dosage and Administration, Drug Interactions, Warnings and Precautions, Adverse Reactions, Clinical Pharmacology, Clinical Studies	Indications and Usage, Warning and Precautions, Clinical Pharmacology, Clinical Studies, Patient Counseling Information	Indications and Usage, Dosage and Administration, Clinical Pharmacology, Clinical Studies
Test	theVysis ALK Break Apart FISH Probe Kit	cobas[®] 4800 BRAF V600 Mutation Test	cobas® EGFR Mutation Test
URL	http://www.abbottmolecular.com/us/products/ oncology/fish/lung-cancer/vysis-lsi-alk-dual-color-break-apart-rearrangement-probe.html	http://molecular.roche.com/assays/Pages/cobas4800BRAFV600MutationTest.aspx	http://molecular.roche.com/assays/Pages/cobasEGFRMutationTest.aspx

For example, in 5-fluorouracil (5-FU), which uses dihydropyrimidine dehydrogenase [DPYD] as a rate-limiting metabolizing enzyme, the presence of a mutation at DPYD*2A will cause partial DPD deficiency, leading to 5-FU toxicity [14-31]. Therefore, other alternatives must be considered, such as changing the drug or changing the dose, depending on the patient's condition. The FDA provides this information on the drug label in the warning section. The Theraguide 5-FU blood test has begun to be used to allow customized management and to minimize the patient's risk for an adverse reaction to 5-FU-related chemotherapies.

In addition, both germline mutations and somatic mutations play roles in drug response. Germline mutations have been used for predicting drug toxicity and efficiency [32] and to develop new or existing therapeutics [33]. Somatic mutations have been used to optimize certain anticancers [32] and to provide accurate drug targets [33]. For example, non-small cell lung cancer sequencing has shown that EGFR tyrosine kinase inhibits sensitivity resistance, which has resulted in the use of routine EGFR sequencing for therapeutic purposes [34].

The application of PG in the drug field begins by targeted identification and discovery, followed by the optimization of a compound, clinical trials, and finally, the launch of a product. Currently, with the huge development in genomic personalized medicine, NGS plays a critical role in designing new therapeutic strategies based on the genetic profile of each individual.

Because cancer is genetically unique, using personalized therapy in cancer is more applicable and should be required. To better understand the concept of designing new therapeutic strategies, DNA sequencing using NGS can detect mutations that will lead to the design of drugs that specifically target these mutations; these drugs are based on genetargeted nucleic acids.

The blocking of apoptosis, cell proliferation, tumor angiogenesis, tumor invasion, tumor suppressor gene dysfunction, increased tumor cell lifespan, DNA synthesis arrest, and drug resistance can all lead to cancer. All of these are carcinogenic effects, the main cause of which is the presence of mutations; therefore, detecting mutant genes will impact the outcome of the therapy (**Figure 2**).

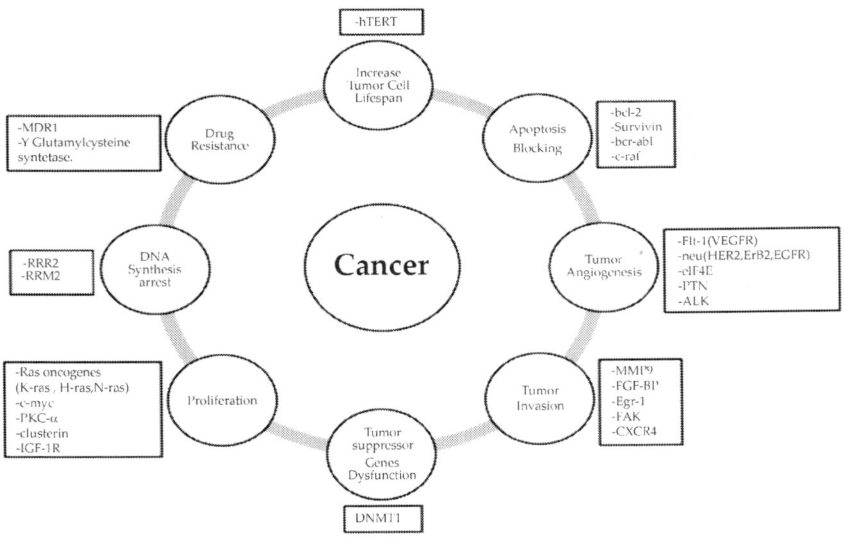

Figure 2: Carcinogenesis Events and Target Genes.

An example of a target gene involved in proliferation is the Ras oncogene family [K-ras, H-ras, N-ras]. The Rasfamily are proteins involved in transmitting signals within cells from tyrosine kinase receptors to the cell nucleus, and they regulate a wide range of processes [35]. The presence of a mutation will cause the loss of the cell's dephosphorylation ability, leading to an increased proliferation rate and cancer cell survival (**Figure 3**) [36,37].

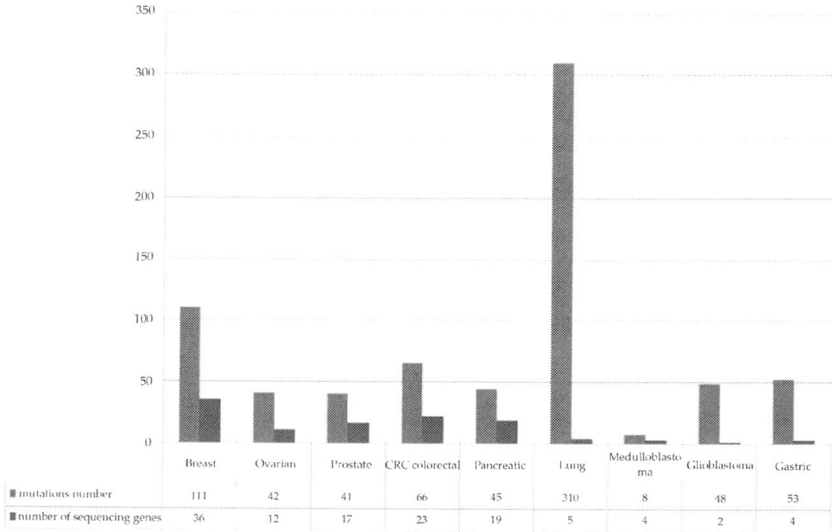

Figure 3: A depiction of the need to apply NGS in cancer showing the number of mutations in different types of cancers and the number of sequenced genes [37].

When using breast cancer as an example, upon comparing the number of mutations (111 mutations) with the number of sequenced genes (36 genes), the results show that there is an absolute need to detect all of the mutations and to use the given information to find the best tools and methods in diagnosis and treatment (**Table 3**).

Table 3: Examining the effect of mutations in more detail by looking at the locations of mutations in different types of cancers and the effects of different anticancer drugs

Cancer Type	mutation	Type/Location of mutation	Effect in anticancer drugs				
			Imatinib	Dasatinib	Nilotinib	Bosutinib	Ponatinib
CML	BCR-ABL1 Fusions	t[9;22][q34.1;q11.21] and other translocations resulting in the BCR-ABL1 fusion gene	Confers sensitivity	Confers sensitivity	Confers Sensitivity	Confers sensitivity	Confers sensitivity
	BCR-ABL1 Y253H	P-loop region of the kinase domain [Exon 4]	Confers reduced sensitivity	Retains sensitivity	Confers reduced sensitivity	Unknown at this time	Unknown at this time
	BCR-ABL1 V299L	SH3 contact region of the BCR-ABL1 kinase domain [Exon 5]	Confers reduced sensitivity	Confers reduced sensitivity	Retains sensitivity	Unknown at this time	Unknown at this time
	BCR-ABL1 T315I	Kinase domain [Exon 6]	Confers reduced sensitivity	Confers reduced sensitivity	Confers reduced sensitivity	Confers reduced sensitivity	Retains sensitivity
BCC	SMO D473H	At the boundary between the 6th transmembrance domain and an extracellular domain	Vismodegib				
			Hypothesized to confer decreased sensitivity				

Breast Cancer	AR Expression		Androgen Receptor Modulators And Antagonists		Estrogen Receptor Agonists/ Antagonists		Aromatase Inhibitors
	ER Expression		Confers increased sensitivity		Confers increased sensitivity		Unknown at this time
	HER2 [ERBB2] Amplification Confers increased sensitivity		Trastuzumab	Ado-TrastuzumabEmtansine	Pertuzumab	Lapatinib	Neratinib
			Confers increased sensitivity	Confers increased sensitivity	Confers increased sensitivity	Confers increased sensitivity	
Thymic Malignancies	KIT E490K	Extra-cellular domain [exon 9]	Imatinib	Sunitinib	Sorafenib	Dasatinib	
			May confer increased sensitivity	May confer increased sensitivity	May confer increased sensitivity	May confer increased sensitivity	
	KIT Y553N	Juxtamembrane domain [exon 11]	Confers sensitivity	Unknown at this time	Unknown at this time	Unknown at this time	
	KIT W557R	Juxtamembrane domain [exon 11]	May confer increased sensitivity	May confer increased sensitivity	May confer increased sensitivity	May confer increased sensitivity	
	KIT V560	Juxtamembrane domain [exon 11]	Confers increased sensitivity	May confer increased sensitivity	Unknown at this time	May confer increased sensitivity	
	KIT P577_D579del	Juxtamembrane domain [exon 11]	Unknown at this time	Unknown at this time	Confers sensitivity	Unknown at this time	
	KIT D820E	Kinase domain [exon 17]	May confer decreased sensitivity	May confer decreased sensitivity	Confers increased sensitivity	May confer increased sensitivity	
Cancer Type	**mutation**	**Type /Location of mutation**	**Implications for Targeted Therapeutics**				

GIST	BRAF	V600E mutation substitution at position 600 in BRAF Kinase domain [exon 15]	Imatinib	Sunitinib
			Confers resistance	Confers resistance
	KIT	Extracellular dimerization motif [Exon 9]	Confers intermediate sensitivity; Higher doses of imatinib[up to 800 mg total daily dose] may be more effective in metastatic disease than 400 mg oral daily	Confers increased sensitivity
		Juxtamembrane domain [Exon 11]	Confers increased sensitivity	Decreased sensitivity to second-line sunitinib; Too few patients have been treated in the imatinib-naïve setting
		Tyrosine kinase 1 [TK1] domain/ATP binding pocket [Exon 13]	Confers sensitivity as a primary mutation; Confers resistance as a secondary mutation	Sensitive in *in vitro* studies; Too few patients in imatinib naïve or second-line settings to determine
		Tyrosine kinase 1 [TK1] domain/ATP binding pocket [exon 14]	Confers resistance as a secondary mutation	Sensitive *in vitro* studies; Too few patients in imatinib naïve or second-line settings to determine
		Tyrosine kinase 2 [TK2] domain and activation loop [exon 17]	Primary mutation sensitive *in vitro*; Confers resistance as a secondary mutation	Confers resistance as secondary mutation; Too few people treated in imatinib-naïve setting to determine activity
	PDGFRA D842V	Tyrosine kinase 2 [TK2] domain [Exon 18]	Confers decreased sensitivity	Confers decreased sensitivity
Bladder Cancer	TSC1 E636fs	frameshift introduction of stop codon into the TSC1 gene [EXON 15]	mTOR inhibitors: Confers increased sensitivity	

IMT	ALK Fusions	Chromosomal rearrangements involving the ALK gene on 2p23	ALK inhibitors: Confers increased sensitivity		
Gastric Cancer	HER2	HER2 amplification	Trastuzumab : Confers increased sensitivity		
NSCLC	ALK Fusions	Chromosomal rearrangements involving the ALK gene on 2p23	ALK TKIs	HSP90 inhibitors	EGFR TKIs
			Confers increased sensitivity	Confers increased sensitivity	Confers decreased sensitivity
	EGFR	Kinase domain [exon 19] Insertion	Unknown at this time	Unknown at this time	Confers increased sensitivity
		Kinase domain [exon 19] DELETION	Unknown at this time	Unknown at this time	Confers increased sensitivity
		Kinase domain [exon 20] Insertion	Unknown at this time	Unknown at this time	Confers decreased sensitivity
	ROS1 Fusions	Chromosomal rearrangements involving the ROS1 gene on 6q22	Crizotinib	Erlotinib/Gefitinib	
			Confers increased sensitivity	Confers decreased sensitivity	

For example, in chronic myelogenous leukemia [CML], which is a type of cancer that affects humans, the presence of BCR-ABL1 fusions at the location t [9;22] [q34.1;q11.21] confers sensitivity to 5 anticancer drugs, whereas the presence of the BCR-ABL1 Y253H mutation in the P-loop region of the kinase domain [exon 4] confers reduced sensitivity to imatinib and nilotinib but retains sensitivity to dasatinib. The effects of ponatinib and bosutinib are unknown. Additional studies and clinical trials are still required to understand their effects.

Targeted Cancer Therapy

Above, we describe how mutations in genes will affect the outcome of therapy and move the science of drug design forward. In cancer research, the development of drugs that target carcinogenesis events is called targeted cancer therapy or rational drug design. The primary goal of these targeted drugs is to interfere with specific molecules involved in cancer without causing any harm to normal cells. Targeted drugs are the cornerstone of precision medicine and the current focus of anticancer discovery and development [38].

Standard chemotherapy works in a non-selective manner because it acts on all rapidly dividing cells, whether they are cancerous or normal, but targeted cancer therapies are designed and synthesized to be delivered to and only react with their target; therefore, they act on specific molecular targets that are associated with cancer. In cytogenetics, targeted therapies are cytostatic [they block tumor cell proliferation], whereas standard chemotherapy agents are cytotoxic [they kill tumor cells] [38].

With the recent advances in genetic medicine and the sequencing of the human genome, the concept of a "druggable genome" has become more applicable and popular [39-41]. A number of approaches are used to identify potential targets, and it is always possible to discover new approaches. One approach is to compare the amount of proteins present in cancer cells with the amount of proteins present in normal cells. An example of a differentially expressed target is the human epidermal growth factor receptor 2 protein [HER-2], which is expressed at high levels in some cancer cells.

Several targeted therapies are directed against HER-2, including trastuzumab [Herceptin®] [38]. In addition, non-small cell lung cancer sequencing has shown the presence of EGFR tyrosine kinase inhibitor sensitivity resistance, which has resulted in the use of routine EGFR sequencing for therapeutic purposes [34]. Recently, a study was published online (23 September 2014) that discussed the use of Mir-34 as a new weapon against cancer [42].

Another approach is to determine whether cancer cells produce mutant [altered] proteins that drive cancer progression. For example, BRAF, which is a cell growth signaling protein, has been found to be present in an altered form [BRAF V600E] in many melanomas. The current approved drug is vemurafenib [Zelboraf®], and it targets the mutant form of the BRAF protein to treat patients with this type of cancer (**Table 4**) [38].

Table 4: Targeted therapies that have been approved by the FDA

FDA-Approved Indication[S]	**Target[S]**	**Targeted Drugs**
Colorectal cancer [KRAS wild type]	EGFR [HER1/ERBB1]	Panitumumab[Vectibix]
Renal cell carcinoma	VEGFR, PDGFR, KIT	Pazopanib[Votrient]
Melanoma	PD-1	Pembrolizumab[Keytruda]
Breast cancer [HER2+]	HER2 [ERBB2/neu]	Pertuzumab[Perjeta]
Chronic myelogenousleuke-mia Acute lymphoblastic leukemia [Philadelphia chromosome positive]	ABL, FGFR1-3, FLT3, VEGFR2	Ponatinib[Iclusig]
Gastric cancer or Gastro-esophageal junction [GEJ] adenocarcinoma	VEGFR2	Ramucirumab[Cyramza]
Colorectal cancer Gastrointestinal stromal tumors	KIT, PDGFRβ, RAF, RET, VEGFR1/2/3	Regorafenib[Stivarga]
Non-Hodgkin's lymphoma Chronic lymphocytic leu-kemia Rheumatoid arthritis Granulomatosis with poly-angiitis	CD20	Rituximab [Rituxan, Mabthera]
Cutaneous T-cell lymphoma Peripheral T-cell lymphoma	HDAC	Romidepsin[Istodax]
Myelofibrosis	JAK1/2	Ruxolitinib[Jakafi]
MulticentricCastleman's disease	IL-6	Siltuximab[Sylvant]

Hepatocellular carcinoma Renal cell carcinoma Thyroid carcinoma	VEGFR, PDGFR, KIT, RAF	Sorafenib[Nexavar]
Renal cell carcinoma	mTOR	Temsirolimus[Torisel]
Rheumatoid arthritis Juvenile idiopathic arthritis	IL-6R	Tocilizumab[Actemra]
Rheumatoid arthritis	JAK3	Tofacitinib[Xeljanz]
Non-Hodgkin's lymphoma	CD20	Tositumomab[Bexxar]
Melanoma [with BRAF V600 mutation]	MEK	Trametinib[Mekinist]
Breast cancer [HER2+] Gastric cancer [HER2+]	HER2 [ERBB2/neu]	Trastuzumab[Herceptin]
Medullary thyroid cancer	EGFR [HER1/ERBB1], RET, VEGFR2	Vandetanib[Caprelsa]
Melanoma [with BRAF V600 mutation]	BRAF	Vemurafenib[Zelboraf]
Basal cell carcinoma	PTCH, Smoothened	Vismodegib[Erivedge]
Cutaneous T-cell lymphoma	HDAC	Vorinostat[Zolinza]
Colorectal cancer	PIGF, VEGFA/B	Ziv-aflibercept[Zaltrap]

The primary methods of targeted cancer therapy include hormone therapies, signal transduction inhibitors, gene expression modulators, apoptosis inducers, angiogenesis inhibitors, immunotherapies, and toxin delivery molecules [38]. Hormone therapies, which are used to treat tumors that are sensitive to hormones, such as breast cancer, uterine cancer and prostate cancer, work by reducing the hormones that stimulate tumor growth or by blocking the hormone receptor. For example, aromatase inhibitors, such as anastrozole, block the formation of estrogen, tamoxifen citrate blocks antiestrogens, bicalutamide blocks androgen receptor antagonists, and abarelix [antagonist] and gosrelin [analog] block GnRH analogs and its antagonists [43].

Kinase inhibitors, which are a new group of anticancers known as signal transduction inhibitors, target the cellular mechanisms of signal transduction and block the activity of molecules that play a role in or participate in these mechanisms. Examples include imatinib, which was the first agent in this group and works as a BCR-ABL kinase inhibitor; dasatinib, which is a multikinase inhibitor; everolimus, which is an mTOR inhibitor; and gefitinib, which is an EGFR tyrosine kinase inhibitor [43]. Gene expression modulators modify the functions of proteins that play a role in controlling gene expression. For example, triplex-forming oligonucleotides [TFO] bind to the major groove of duplex DNA, leading to the creation of a third strand or a triple helix [44,45].

Apoptosis inducers are also used because cancer cells avoid apoptosis. Therefore, the mechanism of action of this group is strategies that result in the death of cancer cells [38].

Angiogenesis inhibitors block the nourishment of the tumor by blocking the growth of new blood vessels or by targeting the molecules responsible for stimulating the growth of new blood vessels. For example, itraconazole inhibits VEGFR phosphorylation [46,47].

Monoclonal antibodies, which are designed to target the proteins expressed in tumor cells, make the cancer cell more visible to the immune system. For example, rituximab, which targets CD20, blocks growth signals; cetuximab, which targets EGFR, stops new blood vessels from forming; bevacizumab, which targets VEGFR, delivers radiation to cancer cells; and ibritumomabtiuxetan, which is labeled with yttrium 90, blocks CD20 [43].

Precision Medicine

Personalized medicine, genomic medicine, stratified medicine, and targeted medicine are concepts that focus on characterizing individual biological profiles by analyzing the genome and using the results to guide medical decisions to provide patients with the best health care. Precision medicine is generally defined as providing the right patient with the right drug at the right time by precisely targeting the molecular pathways that caused the disease. Precision medicine links the multifactorial characteristics of each individual. A unique feature of precision medicine is that it does not depend only on genomic medicine; it also incorporates a patient's lifestyle, non-genomic biological information, environmental parameters and other relevant data [1].

NGS, which has paved the way for precision medicine, has gained many successes in the field of oncology, allowing each individual to have a comprehensive genomic profile. This provides many advantages for the cancer patient. For example, it allows the generation of specific tumor data based on the patient's characteristics, which can then be used later for targeted therapy, genome-based blood tests, biomarkers and other clinical applications.

Roy Chowdhury's report was the first study to apply NGS in personalized oncology. He showed that using such techniques not only improved patient health care but also made the best use of the existing available genetic information. The aim of the study was to discover the practical challenges of using NGS in the field of clinical oncology. The researchers implemented whole-genome sequencing [WGS], targeted whole-exome sequencing [WES] [the coding regions of the genome] of both normal and cancer DNA, and transcriptome sequencing [RNAseq] [transcribed RNAs] for each patient [48].

The results detected many mutation categories [gene expression alterations, structural rearrangements, point mutations, copy number alternations], and the findings were discussed by a multidisciplinary sequencing tumor board [STB], which included a panel of a wide range of specialists, including those working in clinical oncology, cancer genetics, bioinformatics, clinical pathology, social and behavioral sciences and bioethics [49].

Two patients were enrolled in the pilot study, one with melanoma and the other with colorectal cancer, and both were treated unsuccessfully. After determining the final results of their sequencing and taking a deep look into their genetic profiles, the melanoma patient was shown to have HRAS, a CDKN2C structural rearrangement. The sequencing tumor board [STB] suggested using PI3K and MEK inhibitors as a treatment for the patient. Sequencing results in the colorectal cancer patient detected CDK8, an NRAS variation that may be used as a future therapeutic target [48].

Additionally, one of the most successful cases of the use of NGS in clinical practice is an inspiring study that was performed at Washington University and was headed by genetics researchers who performed transcriptome sequencing and whole-genome sequencing [WGS] for a researcher on the team affected by adult acute lymphoblastic leukemia [ALL]. The results showed that the FLT3 gene was highly active in the leukemia cells. The anticancer drug sunitinib, which is approved by the FDA to treat gastrointestinal stromal tumors, renal cell carcinoma, and pancreatic cancer, inhibits FLT3. After using sunitinib, the patient's blood returned to normal (**Table 5**) [50-60].

Table 5: Studies showing the remarkable success of using NGS in precision medicine

Cancer Type	Sample	Major Findings	Findings Importance
Whole Genome sequencing [WG]			
Melanoma	WGS:25	Detect :- -Mutated gene, PREX2 -comprehensive sequencing of human melanoma tumors	PREX2 mutations are biologically related, ectopic expression of mutant PREX2 accelerated tumor formation of immortalized human melanocytes *in vivo*[52].
Breast cancer	WES :103 WGS:22	Detect:- - repeated mutation in CBFB -deletion of RUNX1 -repeated MAGI3-AKT3 fusion	MAGI3-AKT3fusion led to AKT kinase activation which cancelled by ATP-competitive AKT small molecular inhibitor treatment [53].
Whole exome sequencing [WES]			
Uterine serous carcinoma	WES:10	Detect :- Somatic mutions in TP53,PIK3CA,FBXW7,PPP2R1A.	Cyclin E- FBXW7, PI3K, TP53 have a role in the etiology of uterine serous carcinoma [54].
Head and neck squamous cancer	WES:32	Detect :- -mutions in FBXW7, NOTCH1	40% of 28 identified mutations in NOTCH1 were truncate gene product. - NOTCH1 may function as tumor suppressor [55].
Transcriptomic sequencing			
Colon cancer	WGS:2 WES:72 RNA-seq:68	Detect :- -IGF2 over expression -repeted gene fusion RSPO2, RSPO3 .	-RSPO may have a role in Wntsignaling activation and tumerigenesis . -R spondin gene fusion provide new therapeutic opportunities [56].
thyroid cancer cells [TPC-1] and leukemia cells	MicroRNAs:90	defined region upstream of a conserved GXGXXG kinase motif	* Tyrosine kinase [TK] fusions are attractive drug targets in cancers * identified for the first time the genomic fusion sequences of CCDC6-RET in TPC-1 cells and FGFR1OP2-FGFR1 in KG-1 cells [57].
Targeted sequencing	Targeted sequencing	Targeted sequencing	

familial breast cancer	N=12	*identified all 19 distinct BRCA1 and 35 distinct BRCA2 sequence alterations. *detection of BRCA1, BRCA2 variants from introns and untranslated regions	Individuals and families carrying mutations in BRCA1 and BRCA2 [BRCA1/2] have a markedly elevated risk of developing breast and ovarian cancers [58].
Prostate cancer	N=45	-182 cancer associated genes -37 introns of commonly rearranged genes. -novel and actionable BRAF rearranged	Designing targeted assays to detect driving mutations, drug targets, future therapies, biomarkers [59].
Epigenetic sequencing			
Bladder cancer	N=212	- 1,627 hypermethylated promoter targets in the BC cell - VAX1, LMX1A are associated with BC recurrence. - VAX1, KCNV1, TAL1, PPOX1, and CFTR are associated with BC diagnosis.	identified a promising diagnostic marker panel for early non-invasive detection and subsequent BC surveillance [60].
Non-small cell lung Cancer	N=48	- identified 57 differentially methylated regions [DMRs] present in all NSCLC tumors. - hypermethylated DMRs were strongly associated with genes encoding transcriptional regulators. - subtelomeric regions and satellite repeats were hypomethylated in the NSCLC samples. - identified DMRs that were specific to adenocarcinomas and squamous cell carcinomas	provide a resource containing genome-wide DNA methylation maps of NSCLC and their paired lung tissues, and comprehensive lists of known and novel DMRs and associated genes in NSCLC [61].

Biopharmaceutics

One of the major classes of anticancer agents are monoclonal antibodies, which bind to a target and activate the host's immune mechanisms, leading to the killing of the cancer cells by either complete mediated lysis or via killer cells. Monoclonal antibodies can attach to an inactive growth factor receptor on a cancer cell, thus inhibiting survival pathways and promoting apoptosis. Cetuximab is a monoclonal antibody directed against EGFR, erbitux is a well-known anti-EGFR antibody, rituximab lyses B lymphocytes by binding to the calcium channel forming CD20, and trastuzumab binds to HER2.

The importance of NGS in this field lies in generating an antibody library because NGS data can provide a more precise analysis of multiplicity in an antibody library [61]. In addition, NGS can be used in interactome profiling to detect protein interactions with multifunctional enzymes [62], thus providing new potential methods for studying protein-protein interactions and antibody-antigen binding [61].

Polypharmacology

The philosophy of drug design has been converted to the idea of one drug multiple targets rather one drug one target [63-68]. Polypharmacology is still a major challenge in drug development. This technique includes using a single drug that is capable of affecting multiple targets in a specific disease pathway or using a single drug that affects multiple targets connected to multiple disease pathways [69]. Polypharmacology aims to explore drug repurposing, which includes exploring the unidentified off targets of available drugs [70-72]. In regards to anticancers, polypharmacology is considered a primary treatment approach because anticancers are often taken in combination with MOPP, a curative combination prototype that consists of nitrogen mustard, vincristine [Oncovin], procarbazine and prednisone.

The importance of NGS lies in using this approach for targeting, to increase a drug's efficacy and minimize its toxicity. Additionally, NGS can detect a wide variety of rare, actionable mutations that can be used as targets, and polypharmacology aims to find multitargets; therefore, they are complementary to each other.

NGS provides multitargets, and polypharmacology delivers the drug. In addition, NGS can be used for ligand-based predictions, target-based predictions, and phenotype-based predictions [including transcriptomic-based methods, proteomics-based approaches, and drug target identification [73].

Toxgnostics

Most anticancers are toxic and cause severe side effects, such as cisplatin, which causes nephrotoxicity and ototoxicity; temozolomide, which causes genotoxicity, fetotoxicity, and teratogenicity; and 6-mercaptopurine, which is dose-dependent and causes hepatotoxicity. These side effects have led to the field of toxgnostics in cancer medicine. Toxgnostics, which is generally defined as a systemic study, focuses on studying genetic toxicity predictors that are related to and caused by anticancers [74].

NGS revolutionized our ability to predict anticancer toxicity, determine the causes of toxicity and detect toxgnostic variants. NGS can also predict the toxicity level in each patient based on the genetic profile and the number of people who may be responsive to a specific drug. Toxgnostics aims to find the relationship between a plasma drug concentration profile and the toxicity level and to create a personalized treatment plan for each patient, which may require elective dose adjustment for a select drug [74].

Vaccinology

NGS can be utilized in the initial development stages 3 of a vaccine to evaluate its efficiency and safety [75]. Due to the accuracy of NGS and its ability to detect sequence variants in each individual, NGS can provide more reasons and solutions to help understand why some people may face side effects after receiving vaccines, but others may not [61].

Pharmacoepidemiology

Pharmacoepidemiology, which uses variations in study designs to determine the effects of drug treatments on populations and clinical situations [76]. NGS has been successfully used in this field because epigenetics plays a role in tumorigenesis and paves the way for drug discovery. The two important mechanisms of epigenetics are DNA methylation and histone modification. For example, epigenetics can be used for the following:

- epigenetic-based biomarkers [for colorectal cancer, septin 9 is used to block methylation] [77].
- epigenetic-based drugs [for treatment of cutaneous T-cell lymphoma, vorinostat, a histone deacetylase inhibitor from Merck, is used [78,79].

CONCLUSION

Next-generation sequencing has proven to be the cornerstone of advanced research. This includes not only cancer research but also all areas of human genetics. By improving the speed and precision of the results, designing new therapeutic strategies can be achieved to improve the outcome of cancer treatments and save as many lives as possible.

REFERENCES

1. Servant N,Roméjon J,Gestraud P, La Rosa P,Lucotte G, et al. (2014) Bioinformatics for precision medicine in oncology: principles and application to the SHIVA clinical trial. Front Genet 5: 152.
2. Weinberg RA (2013) The Biology of Cancer.(2ndedn).
3. Neidle S (2013) Cancer Drug Design and Discovery.Elsevier Inc.
4. Wu W, Choudhry H (2013) Next Generation Sequencing in Cancer Research. Springer.
5. Gates RA, Regina MF(2008) Oncology Nursing Secrets. (3rdedn).
6. EzpeletaNR,Hackenberg M, Aransay AM (2012) Bioinformatics for High Throughput Sequencing.
7. Ozretia L,Heukamp LC, Odenthal M, Buettner R (2012) The role of molecular diagnostics in cancer diagnosis and treatment. Onkologie 35 Suppl 1: 8-12.
8. Guan YF, Li GR, Wang RJ, Yi YT, Yang L, et al. (2012) Application of next-generation sequencing in clinical oncology to advance personalized treatment of cancer. Chin J Cancer 31: 463-470.
9. Cronin M, Ross JS (2011) Comprehensive next-generation cancer genome sequencing in the era of targeted therapy and personalized oncology. Biomark Med 5: 293-305.
10. http://www.cancer.org/treatment/understandingyourdiagnosis/examsandtestdescriptions/tumormarkers/#
11. Wu JT (2002) Circulating Tumor Markers of the New Millennium: Target Therapy, Early Detection, and Prognosis, Clinical chemistry. (1stedn), AmerAssn for Clinical Chemistry.
12. Wu JT (1999) Review of circulating tumor markers: from enzyme, carcinoembryonic protein to oncogene and suppressor gene. Ann Clin Lab Sci 29: 106-111.
13. Nishant T, Bindu HK, Kumar SD, Kumar AR (2012) Pharmacogenomics- Personalized Treatment of Cancer, Diabetes and

Cardiovascular Diseases. J PharmacogenomPharmacoproteomics 3: 107.

14. Lee A, Ezzeldin H, Fourie J, Diasio R (2004) Dihydropyrimidine dehydrogenase deficiency: impact of pharmacogenetics on 5-fluorouracil therapy. Clinical advances in hematology &oncology: H&O 2: 527-532.
15. Dancey JE,Bedard PL, Onetto N, Hudson TJ (2012) The genetic basis for cancer treatment decisions. Cell 148: 409-420.
16. Choi YL, Soda M, Yamashita Y, Ueno T, Takashima J, et al. (2010) EML4-ALK mutations in lung cancer that confer resistance to ALK inhibitors. N Engl J Med 363: 1734-1739.
17. Sasaki T, Okuda K, Zheng W, Butrynski J, Capelletti M, et al. (2010) The neuroblastoma-associated F1174L ALK mutation causes resistance to an ALK kinase inhibitor in ALK-translocated cancers. Cancer Res 70: 10038-10043.
18. Inamura K, Takeuchi K, Togashi Y, Hatano S, Ninomiya H, et al. (2009) EML4-ALK lung cancers are characterized by rare other mutations, a TTF-1 cell lineage, an acinar histology, and young onset. Mod Pathol 22:508-515.
19. Koivunen JP,Mermel C, Zejnullahu K, Murphy C, Lifshits E, et al. (2008) EML4-ALK fusion gene and efficacy of an ALK kinase inhibitor in lung cancer. Clin Cancer Res 14: 4275-4283.
20. Shaw AT,Yeap BY, Mino-Kenudson M, Digumarthy SR, Costa DB, et al. (2009) Clinical features and outcome of patients with non-small-cell lung cancer who harbor EML4-ALK. J ClinOncol 27: 4247-4253.
21. Soda M, Takada S, Takeuchi K, Choi YL, Enomoto M, et al. (2008) A mouse model for EML4-ALK-positive lung cancer. ProcNatlAcadSci U S A 105: 19893-19897.
22. Flaherty KT,Puzanov I, Kim KB, Ribas A, McArthur GA, et al. (2010) Inhibition of mutated, activated BRAF in metastatic melanoma. N Engl J Med 363: 809-819.
23. Johannessen CM, Boehm JS, Kim SY, Thomas SR, Wardwell L, et al. (2010) COT drives resistance to RAF inhibition through MAP kinase pathway reactivation. Nature 468: 968-972.

24. Nazarian R, Shi H, Wang Q, Kong X, Koya RC, et al. (2010) Melanomas acquire resistance to B-RAF(V600E) inhibition by RTK or N-RAS upregulation. Nature 468: 973-977.
25. Pao W, Miller V, Zakowski M, Doherty J, Politi K, et al. (2004) EGF receptor gene mutations are common in lung cancers from "never smokers" and are associated with sensitivity of tumors to gefitinib and erlotinib. ProcNatlAcadSci USA 101: 13306-13311.
26. Paez JG,Jänne PA, Lee JC, Tracy S, Greulich H, et al. (2004) EGFR mutations in lung cancer: correlation with clinical response to gefitinib therapy. Science 304: 1497-1500.
27. Lynch TJ, Bell DW, Sordella R, Gurubhagavatula S, Okimoto RA, et al. (2004) Activating mutations in the epidermal growth factor receptor underlying responsiveness of non-small-cell lung cancer to gefitinib. N Engl J Med 350: 2129-2139.
28. Riely GJ, Politi KA, Miller VA, Pao W (2006) Update on epidermal growth factor receptor mutations in non-small cell lung cancer. Clin Cancer Res 12: 7232-7241.
29. Toschi L,Cappuzzo F (2007) Understanding the new genetics of responsiveness to epidermal growth factor receptor tyrosine kinase inhibitors. Oncologist 12: 211-220.
30. Pao W, Miller VA, Politi KA, Riely GJ, Somwar R, et al. (2005) Acquired resistance of lung adenocarcinomas to gefitinib or erlotinib is associated with a second mutation in the EGFR kinase domain. PLoS Med 2: e73.
31. Yun CH,Mengwasser KE, Toms AV, Woo MS, Greulich H, et al. (2008) The T790M mutation in EGFR kinase causes drug resistance by increasing the affinity for ATP. ProcNatlAcadSci U S A 105: 2070-2075.
32. Weng L, Zhang L, Peng Y, Huang RS (2013) Pharmacogenetics and pharmacogenomics: a bridge to individualized cancer therapy. Pharmacogenomics 14: 315-324.
33. Smith SA, French T, Hollingsworth SJ (2014) The impact of germline mutations on targeted therapy. J Pathol 232: 230-243.
34. McLeod HL (2013) Cancer pharmacogenomics: early promise, but concerted effort needed. Science 339: 1563-1566.

35. Barbacid M (1987) ras genes. Annu Rev Biochem 56: 779-827.
36. Campbell SL,Khosravi-Far R, Rossman KL, Clark GJ, Der CJ (1998) Increasing complexity of Ras signaling. Oncogene 17: 1395-1413.
37. Vogelstein B, Papadopoulos N, Velculescu VE, Zhou S, Diaz LA Jr, et al. (2013) Cancer genome landscapes. Science 339: 1546-1558.
38. http://www.cancer.gov/cancertopics/treatment/types/targeted-therapies/targeted-therapies-fact-sheet
39. Patel MN,Halling-Brown MD, Tym JE, Workman P, Al-Lazikani B (2013) Objective assessment of cancer genes for drug discovery. Nat Rev Drug Discov 12: 35-50.
40. Hopkins AL, Groom CR (2002) The druggable genome. Nat Rev Drug Discov 1: 727-730.
41. Overington JP, Al-Lazikani B, Hopkins AL (2006) How many drug targets are there? Nat Rev Drug Discov 5: 993-996.
42. Misso G, Di Martino MT, De Rosa G,Farooqi AA, Lombardi A, et al. (2014) Mir-34: a new weapon against cancer? MolTher Nucleic Acids 3: e194.
43. Cohen V (2012) Basic Concepts in Pharmacology: What You Need to Know for Each Drug Class, 4th Edition. Ann Pharmacother .
44. Sargent RG, Kim S, Gruenert DC (2011) Oligo/polynucleotide-based gene modification: strategies and therapeutic potential. Oligonucleotides 21: 55-75.
45. Uil TG,Haisma HJ, Rots MG (2003) Therapeutic modulation of endogenous gene function by agents with designed DNA-sequence specificities. Nucleic Acids Res 31: 6064-6078.
46. Aftab BT,Dobromilskaya I, Liu JO, Rudin CM (2011) Itraconazole inhibits angiogenesis and tumor growth in non-small cell lung cancer. Cancer Res 71: 6764-6772.
47. Chong CR,Xu J, Lu J, Bhat S, Sullivan DJ Jr, et al. (2007) Inhibition of angiogenesis by the antifungal drug itraconazole. ACS ChemBiol 2: 263-270.
48. Roychowdhury S,Iyer MK, Robinson DR, Lonigro RJ, Wu YM, et al. (2011) Personalized oncology through integrative high-throughput sequencing: a pilot study. SciTransl Med 3: 111ra121.

49. Corless CL (2011) Medicine. Personalized cancer diagnostics. Science 334: 1217-1218.

50. http://pharmaceuticalintelligence.com/2012/07/09/sunitinib-brings-adult-all-to-remission-rna-sequencing/

51. Berger MF,Hodis E, Heffernan TP, Deribe YL, Lawrence MS, et al. (2012) Melanoma genome sequencing reveals frequent PREX2 mutations. Nature 485: 502-506.

52. Banerji S,Cibulskis K, Rangel-Escareno C, Brown KK, Carter SL, et al. (2012) Sequence analysis of mutations and translocations across breast cancer subtypes. Nature 486: 405-409.

53. Kuhn E, Wu RC, Guan B, Wu G, Zhang J, et al. (2012) Identification of molecular pathway aberrations in uterine serous carcinoma by genome-wide analyses. J Natl Cancer Inst 104: 1503-1513.

54. Agrawal N, Frederick MJ, Pickering CR, Bettegowda C, Chang K, et al. (2011) Exome sequencing of head and neck squamous cell carcinoma reveals inactivating mutations in NOTCH1. Science 333: 1154-1157.

55. Seshagiri S,Stawiski EW, Durinck S, Modrusan Z, Storm EE, et al. (2012) Recurrent R-spondin fusions in colon cancer. Nature 488: 660-664.

56. Chmielecki J, Peifer M, Jia P, Socci ND, Hutchinson K, et al. (2010) Targeted next-generation sequencing of DNA regions proximal to a conserved GXGXXG signaling motif enables systematic discovery of tyrosine kinase fusions in cancer. Nucleic Acids Res 38: 6985-6996.

57. Ozcelik H, Shi X, Chang MC, Tram E, Vlasschaert M, et al. (2012) Long-range PCR and next-generation sequencing of BRCA1 and BRCA2 in breast cancer. J MolDiagn 14: 467-475.

58. Beltran H, Yelensky R, Frampton GM, Park K, Downing SR, et al. (2013) Targeted next-generation sequencing of advanced prostate cancer identifies potential therapeutic targets and disease heterogeneity. EurUrol 63: 920-926.

59. Zhao Y,Guo S, Sun J, Huang Z, Zhu T, et al. (2012) Methylcap-seq reveals novel DNA methylation markers for the diagnosis and recurrence prediction of bladder cancer in a Chinese population. PLoS One 7: e35175.

60. Carvalho RH,Haberle V, Hou J, van Gent T, Thongjuea S, et al. (2012) Genome-wide DNA methylation profiling of non-small cell lung carcinomas. Epigenetics Chromatin 5: 9.

61. Woollard PM, Mehta NA, Vamathevan JJ, Van Horn S, Bonde BK, et al. (2011) The application of next-generation sequencing technologies to drug discovery and development. Drug Discov Today 16: 512-519.

62. Di Niro R, Sulic AM, Mignone F, D'Angelo S, Bordoni R, et al. (2010) Rapid interactome profiling by massive sequencing. Nucleic Acids Res 38: e110.

63. Hopkins AL (2008) Network pharmacology: the next paradigm in drug discovery. Nat ChemBiol 4: 682-690.

64. Hopkins AL (2009) Drug discovery: Predicting promiscuity. Nature 462: 167-168.

65. Apsel B, Blair JA, Gonzalez B, Nazif TM, Feldman ME, et al. (2008) Targeted polypharmacology: discovery of dual inhibitors of tyrosine and phosphoinositide kinases. Nat ChemBiol 4: 691-699.

66. Simon Z,Peragovics A, Vigh-Smeller M, Csukly G, Tombor L, et al. (2012) Drug effect prediction by polypharmacology-based interaction profiling. J ChemInf Model 52: 134-145.

67. Brianso F, Carrascosa MC, Oprea TI, Mestres J (2011) Cross-pharmacology analysis of G protein-coupled receptors. Curr Top Med Chem 11: 1956-1963.

68. Paolini GV,Shapland RH, van Hoorn WP, Mason JS, Hopkins AL (2006) Global mapping of pharmacological space. Nat Biotechnol 24: 805-815.

69. Reddy AS, Zhang S (2013) Polypharmacology: drug discovery for the future. Expert Rev ClinPharmacol 6: 41-47.

70. Oprea TI,Mestres J (2012) Drug repurposing: far beyond new targets for old drugs. AAPS J 14: 759-763.

71. Oprea TI, Nielsen SK, Ursu O, Yang JJ, Taboureau O, etal. (2011) Associating Drugs, Targets and Clinical Outcomes into an Integrated Network Affords a New Platform for Computer-Aided Drug Repurposing. Mol Inform 30: 100-111.

72. Achenbach J,Tiikkainen P, Franke L, Proschak E (2011)

Computational tools for polypharmacology and repurposing. Future Med Chem 3: 961-968.

73. Tang J, Aittokallio T (2014) Network pharmacology strategies toward multi-target anticancer therapies: from computational models to experimental design principles. Curr Pharm Des 20: 23-36.
74. Church D, Kerr R, Domingo E,Rosmarin D,Palles C, et al. (2014) 'Toxgnostics': an unmet need in cancer medicine. Nat Rev Cancer 14: 440-445.
75. Victoria JG, Wang C, Jones MS, Jaing C, McLoughlin K, et al. (2010) Viral nucleic acids in live-attenuated vaccines: detection of minority variants and an adventitious virus. J Virol 84: 6033-6040.
76. Freedman AN, Sansbury LB, Figg WD, Potosky AL, Weiss Smith SR, et al. (2010) Cancer pharmacogenomics and pharmacoepidemiology: setting a research agenda to accelerate translation. J Natl Cancer Inst 102:1698-1705.
77. Grützmann R, Molnar B, Pilarsky C, Habermann JK, Schlag PM, et al. (2008) Sensitive detection of colorectal cancer in peripheral blood by septin 9 DNA methylation assay. PLoS One 3: e3759.
78. Delmore JE,Issa GC, Lemieux ME, Rahl PB, Shi J, et al. (2011) BET bromodomain inhibition as a therapeutic strategy to target c-Myc. Cell 146: 904-917.
79. Kaiser J (2010) Epigenetic drugs take on cancer. Science 330: 576-578.

CHAPTER 9

CLINICAL INTERPRETATION AND IMPLICATIONS OF WHOLE GENOME SEQUENCING

Hubert E Blum[1] and Konrad Oexle[2]

[1]Department of Medicine, University Medical Center Freiburg, Germany
[2]Institute of Human Genetics, Technical University Munich, Germany

WHOLE GENOME SEQUENCING

In 2013, 60 years after the discovery of the double-helical structure of the DNA by James Watson and Francis Crick and 10 years after the completion of the sequence of the human genome, the first market authorization of a high throughput ('next-generation') sequencer (Illumina MiSeqDx) was granted by the American Food and Drug Administration. Next-generation sequencing (NGS) allows to effectively performing whole genome sequencing (WGS) that may lead to the introduction of new genome-based analyses into clinical practice [1]. WGS is one of very few recent developments in medicine which are characterized by the attributes 'better', 'faster' and 'cheaper'. With the most recent development of NGS technologies (i.e., Illumina HiSeqX-Ten), complete sequencing of the human genome is feasible within a day for less than US $ 1'000 (not including costs for curation). Several companies offer NGS solutions and compete for a market share. Technologies differ with respect to their miniaturization (e.g., beads, DNA clusters, DNA nanoballs, single molecule sequencing, nanopores). Most still involve sequence-specific nucleic acid hybridization and amplification. Differences exist in the average reading length as well as in the rate, reproducibility, and cost of sequencing. NGS always involves non-directed mass sequencing. Restriction of sequencing on a genomic region, a panel of genes, or the exome (all protein-coding exons), for instance, can be achieved by a capturing tool based on hybridizing oligonucleotides that define the sequence area. The application of NGS for the analysis of RNAs ('transcriptome') or epigenetic modifications ('methylome') is straightforward. For biomedical scientists as well as for clinicians NGS promises nearly unlimited possibilities to search for genetic variants with clinical relevance. This includes the detection of constitutional as well as of somatic mutations as causes of monogenic diseases, the genome-wide assignment of risk scores for polygenic diseases, the identification of tumor- or metastasis-specific mutations in individual tumor patients as well as the determination of gene expression profiles (signatures') with prognostic and/or therapeutic relevance [2,3]. Another example is pharmacogenomics that may contribute to a personalized therapy, allowing choosing for the individual patient the appropriate drug and/or the appropriate dose.

With simplified access to and the reduced cost of WGS or Whole Exome Sequencing (WES), commercialization of these analyses as well as their increasing application in medicine is to be expected. WGS and WES have not only the potential to answer specific disease-related questions [4] but at the same time may identify clinically relevant genetic variants that are not related to the clinical indication for DNA sequencing. From the identification of unexpected variants, ethical issues may arise with respect to their communication to the patient, parents or others [5]. While these questions are very important, our comment primarily focuses on the utility, reliability, resource needs and other aspects of WGS/WES.

In a recently published study, the WGS technology was evaluated in 12 healthy volunteers [4], using the Illumina/Solexa sequencing principle (paired-end reversible terminator massively parallel sequencing of microscopically amplified DNA clusters) on an HiSeq 2000 platform and, for comparison, the nanoarray technology of Complete Genomics Inc. (rolling circle amplification of DNA fragments followed by massively parallel sequencing of arrayed DNA nanoballs using probe-anchor ligation). Initial quality control and genotype annotation followed standardized procedures. The sequence data obtained were scrutinized for small sets of variants associated with cardiovascular risk or clinical drug response as well as a large set of variants likely or known to be involved in inherited diseases. The latter set was subjected to a two-step evaluation process. At first, an automatedalgorithm (STMP, 'Sequence To Medical Phenotypes') prioritized potentially pathogenic, previously known or novel variants according to the usual indicators, such as allele frequency (more common variants are less likely to be pathogenic), functional class, and evidence of evolutionary constraints. The STMP software selected 90-127 variants per individual which included possibly disease-causing variants and variants with the possible implication of a carrier status (8-18 per individual). These variants then were curated by a multidisciplinary team of specialists including genetic counselors, physician/information technology specialists, and molecular pathologists. The study addressed the concordance of the two sequencing platforms (99-100% for single nucleotide variants, 53-59% for insertion/deletion variants), coverage (i.e., readability) of disease-causing genes (incomplete in 10-19% of the genes), validity of the automated evaluation process, i. e., the final number of reportable (2-6 per individual) and 'actionable' variants [5], respectively (1 individual with a clearly actionable result, 4 individuals with variants that are disease-causing according to the 'Human Gene Mutation Database', HGMD, but were reclassified by the curation team as 'reportable with unclear significance'), the correlation between the curators (moderate), the resources needed for sequencing and curation (median time 54 minutes for each curated variant, resulting in median cost of about US $ 15'000 for sequencing, including curation), and the cost of clinical follow-up (less than US $ 1'000 per individual). The authors arrived at several conclusions that will be discussed in the following in the context of clinical aspects of WGS/WES analyses.

1. The authors concluded that incomplete coverage of a considerable fraction of disease-causing genes reduces the sensitivity of WGS/WES analyses. Depending on the NGS platform used, 10-19% of disease-causing genes did not meet the threshold of base

readability accepted by the authors. Without giving a detailed definition of their accepted threshold, it seems to be set at 1%. Thus, the sequences of most of the disease-causing genes were in fact readable. While a readability of 99% of a gene is a problem in single gene analysis, it is less problematic in genome-wide analyses where sensitivity correlates with disease-associated and not with gene-associated variants. Further, the overall coverage of the WGS analyses can be increased, thereby closing some sequence gaps while at the same time increasing the cost of sequencing.

2. The authors observed a low reproducibility of the identification of insertions and deletions between the platforms (<33%) but did not assess the relation between reproducibility and size of the variants. The low reproducibility is indeed very important because the insertion/ deletion variants are likely to cause a loss of gene function. The technological problem underlying the low reproducibility still awaits a satisfactory solution. While the detection of insertions and deletions in WGS indeed may not yet be reliable enough, the concomitant moderate reduction in WGS sensitivity does not preclude its diagnostic application in patients. In fact, there are classes of mutations, e. g., repeat expansions, that cause Huntington's and various other neurological diseases, which currently cannot be detected by NGS technologies. To expect a 100% sensitivity of WGS analyses is and probably will remain unrealistic.

3. The automated sequence interpretation STMP software (see above) results in a high number of seemingly relevant variants (median 108 per individual) that require intensive curation by specialists but still leave an uncertainty in many cases. With this observation, the authors confirmed the findings of previous larger studies [6-9]. The lack of specificity in automated interpretation may be due to (i) false annotations in data bases, (ii) annotation of scientific errors resulting from different analytical problems, ascertainment biases in penetrance estimates, ethnicity effects, or irreproducibility of rare mutations, (iii) insufficient sophistication and power of interpretation algorithms in correctly predicting the

effects of (novel) variants on an individual genetic background, and (iv) to the unclear definition of what is a disease [9,10]. The authors correctly concluded that the lack of specificity impairs the validity of reporting incidental findings of genome-wide analyses. It has to be noted, however, that the above-mentioned limitations in principle also affect single gene analyses by conventional Sanger sequencing. They rarely cause mistakes of these analyses because the probability of identifying a disease-causing variant is usually high if single gene analyses are performed based on a clinical finding. For the same reason, the specificity of WGS/WES analyses is improved if the search area can be restricted by clinical information [3] or by focusing on de novo mutations. Nevertheless, novel guidelines for demonstrating a causal relationship with sequence variants [10] must be followed in order to improve the interpretation of WGS/WES findings in individual patients as well as in the general population.

4. In their 12 healthy study participants the authors identified only 1 variant (frame-shifting deletion of BRCA1) that resulted in a therapeutic or prophylactic action. In a previous exome study [8], including 500 healthy Europeans, the prevalence of clinically relevant genetic variants in adults was 3.4% (determined in a subset of 52/56 'actionable' genes defined by the American College of Medical Genetics and Genomics [5]). Thus, the yield of WGS in unselected individuals is low. On the other hand, 11/12 study participants had 1 or more variants that affected the use or the dosing of a drug.

5. Importantly, significant time is required (about 100 hours per individual) for professional curation after automated evaluation of WGS data by the STMP software. This greatly contributes to the estimated overall cost of about US $ 15'000 per WGS. In principle, there are potentials for process optimization. For instance, 574/1'300 variants identified by the STMP software for professional curation, were listed in the HGMD data base. Among these, 441 were not considered 'disease causing' or 'likely disease causing'. The elimination of these 441 variants would have significantly reduced the time for professional curation without much loss in sensitivity [4]. Overall, the study clearly demon-

strated that cost of sequencing is becoming a less important while data curation and interpretation is the major factor of the overall cost of WGS analyses.

In summary, the WGS study [4] confirmed known problems of WGS/WES analyses in unselected/healthy individuals: limited sensitivity for some classes of mutations such as structural variants, high false-positive rates in automated interpretation, high time requirement for professional curation, and a relatively low impact on medical care. In part, this also affects the diagnostic application of WGS/WES in patients [3] as long as there is no *in vitro* or *in silico* patient- or disease-specific restriction of the sequence analyses. Therefore, there is a great need for the development of restriction strategies that increase the probability of the detection of pathogenic variants, for the improvement of the molecular understanding of genetic effects, interactions and penetrance [9], for further sophistication of effect predicting software, for improvements and quality control of public data bases of mutations and other genetic variants associated with human diseases, for incentives for sharing genotypic and phenotypic data as well as for large-scale NGS genotyping of population cohorts [10].

REFERENCES

1. Collins FS, Hamburg MA (2013) First FDA authorization for next-generation sequencer. N Engl J Med 369: 2369-2371.
2. Kandoth C, McLellan MD, Vandin F, Ye K, Niu B, et al. (2013) Mutational landscape and significance across 12 major cancer types. Nature 502: 333-339.
3. Yang Y, Muzny DM, Reid JG, Bainbridge MN, Willis A, et al. (2013) Clinical whole-exome sequencing for the diagnosis of mendelian disorders. N Engl J Med 369: 1502-1511.
4. Dewey FE, Grove ME, Pan C, Goldstein BA, Bernstein JA, et al. (2014) Clinical interpretation and implications of whole-genome sequencing. JAMA 311: 1035-1045.
5. Green RC, Berg JS, Grody WW, Kalia SS, Korf BR, et al. (2013) ACMG recommendations for reporting of incidental findings in clinical exome and genome sequencing. Genet Med 15: 565-574.
6. MacArthur DG, Balasubramanian S, Frankish A, Huang N, Morris

J, et al. (2012) A systematic survey of loss-of-function variants in human protein-coding genes. Science 335: 823-828.

7. Xue Y, Chen Y, Ayub Q, Huang N, Ball EV, et al. (2012) Deleterious- and disease-allele prevalence in healthy individuals: insights from current predictions, mutation databases, and population-scale resequencing. Am J Hum Genet 91: 1022-1032.
8. Dorschner MO, Amendola LM, Turner EH, Robertson PD, Shirts BH, et al. (2013) Actionable, pathogenic incidental findings in 1,000 participants' exomes. Am J Hum Genet 93: 631-640.
9. Cooper DN, Krawczak M, Polychronakos C, Tyler-Smith C, Kehrer-Sawatzki H (2013) Where genotype is not predictive of phenotype: towards an understanding of the molecular basis of reduced penetrance in human inherited disease. Hum Genet 132: 1077-1130.
10. MacArthur DG, Manolio TA, Dimmock DP, Rehm HL, Shendure J, et al. (2014) Guidelines for investigating causality of sequence variants in human disease. Nature 508: 469-476.

CHAPTER 10

ACCURACY OF NEXT GENERATION SEQUENCING PLATFORMS

Edward J Fox,[1] Kate S Reid-Bayliss,[1] Mary J Emond,[2] and Lawrence A Loeb[1]

[1]Departments of Pathology and Biochemistry, University of Washington, USA

[2]Department of Biostatistics, University of Washington, USA

ABSTRACT

Next-generation DNA sequencing has revolutionized genomic studies and is driving the implementation of precision diagnostics. The ability of these technologies to disentangle sequence heterogeneity, however, is limited by their relatively high error rates. A Several single molecule barcoding strategies have been propose to reduce the overall error frequency. A Duplex Sequencing additionally exploits the fact that DNA is double-strand, with one strand reciprocally encoding the sequence information of its complement, and can eliminate nearly all sequencing errors by comparing the sequence of individually tagged amplicons derived from one strand of DNA with that of its complementary strand.

This method reduces errors to fewer than one per ten million nucleotides sequenced.

Keywords: Next-generation DNA sequencing, Precision medicine, Accuracy, Duplex sequencing

INTRODUCTION

Mutation drives evolution and underlies many diseases, most prominently cancer [1]. Of the newly developed genomic technologies, next-generation DNA sequencing (NGS), in particular, has revolutionized the scale of study of biological systems [2] and has already started to enter the clinic where it is expected to enable a more personalized approach to patient care [3]. Unlike conventional sequencing techniques, which simply report the average genotype of an aggregate of molecules, NGS digitally tabulates the sequence of individual DNA fragments, thereby offering the unique ability to detect minor variants within heterogeneous mixtures [4]. Already, NGS has been used to characterize exceptional diversity within microbial [5,6], viral [7-9], and tumor cell populations [10-12], and many low frequency, drug-resistant variants of therapeutic importance have been identified [13,14]. NGS has also revealed previously underappreciated intra-organismal mosaicism in both the nuclear [15] and mitochondrial genomes [16]. This somatic heterogeneity, along with that underlying adaptive immunity [17], is an important factor in determining the phenotypic variability of disease.

In theory, DNA subpopulations of any size should be detectable via 'deep sequencing' of a sufficient number of molecules. However, a fundamental limitation of standard NGS is the high frequency with which bases are scored incorrectly due to artifacts introduced during sample preparation and sequencing [18]. For example, amplification bias during PCR of heterogeneous mixtures can result in skewed populations [19]. Additionally, polymerase mistakes, such as base misincorporations and rearrangements due to template switching, can result in incorrect variant calls. Furthermore, errors arise during cluster amplification, sequencing cycles, and image analysis result in approximately 0.1–1% of bases being called incorrectly (Table 1).

Table 1: Comparison of the primary error frequencies of DNA sequencing platforms and tag-based error correction methodologies

Commercial Platform	**Most Frequent Error Type**	**Error Frequency**
Capillary sequencing	single nucleotide substitutions	10^{-1}
454 GS Junior	Deletions	10^{-2}
PacBio RS	CG deletions	10^{-2}
Ion Torrent PGM	Short deletions	10^{-2}
Solid	A-T bias	2×10^{-2}
Illumina MiSeq	single nucleotide substitutions	10^{-3}
Illumina HiSeq2000	single nucleotide substitutions	10^{-3}
Tag-based methods		
SafeSeq	single nucleotide substitutions	1.4×10^{-5}
CircleSeq	single nucleotide substitutions	7.6×10^{-6}
Duplex Sequencing	Single nucleotide substitutions	5×10^{-8}

For a genetically homogenous sample, the effects of these base miscalls can be mitigated by establishing a consensus sequence from high-coverage sequencing reads.

However, when rare genetic variants are sought, this base call error frequency presents a profound barrier and has limited the use of deep sequencing in a variety fields that require the highly accurate disentangling of subpopulations within complex (heterogeneous or mixed) biological samples, including metagenomics [20,21], forensics [22], paleogenomics [23] and human genetics [4,24]. Furthermore, for many applications, such as the prenatal screening for fetal aneuploidy [25,26], detection of circulating tumor DNA [27], and monitoring response to chemotherapy with nucleic acid-based serum biomarkers [28], a level of detection well below 1 in 10,000 is highly desirable; unfortunately, the high frequency of erroneous base calls inherent to standard NGS imposes a practical limit of detection of approximately 1 in 100. These technical shortcomings have also limited the elucidation of mechanism by which genomes, and DNA itself, have evolved [29-31], where bioinformatics analyses have been used to reconstruct phylogenetic relationships [32-35].

Although biochemical protocols [36-39] and bioinformatics [10,40-43] have improved sequencing accuracy, the ability to confidently resolve subpopulations below 1% has remained problematic [44]. Laird and colleagues demonstrated that it was possible to significantly reduce the frequency of variant miscalls by covalently linking individual DNA molecules to unique tags prior to amplification [45,46]. This 'barcoding' technique allows many artifactual variations in the sequence to be identified as due to technical error [47-52], as all amplicons derived from a particular individual starting molecule carry the same unique specific tag and can, thus, be collapsed to a consensus sequence representing that of the original DNA strand. An alternative to single-stranded tagging based on shear-points is the circle sequencing methodology developed by Lou et al., which utilizes the strand-displacement activity of Phi29's DNA polymerase to generate multiple copies of circularized DNA molecules in tandem prior to amplification [53]. After sequencing, these linked copies are collapsed to a consensus sequence, thereby eliminating many artifactual errors. Though significant improvements, these single-strand approaches all (Table 1) still exhibit error frequencies greater than the estimated frequency of variation of many biological systems. The mutation rate of normal cells, for example, is estimated to range from 10^{-9} to 10^{-11} mutations/per nucleotide/per cell division [54,55].

Schmitt et al., highlighted a conceptual shortcoming of initial tag-based methods, and of next-generation sequencing platforms in general, in that use is made of sequence data derived from a single strand of DNA [56]. As a consequence, artifactual variants introduced during the initial rounds of PCR amplification become fixed and are indistinguishable from true variants, since the sequence information of the complementary strand is not taken into account. Damage to DNA from oxidative cellular processes, or generated ex vivo during tissue processing and DNA extraction [57,58], is a particular concern, as such damage can result in frequent copying errors by DNA polymerases. For example, the most thoroughly studied DNA lesion arising from oxidative damage, 8-oxoguanine, incorrectly pairs with adenine during copying with an overall efficiency greater than that of correct pairing with cytosine, and can, thus, contribute a large frequency of artifactual G:C→T:A mutations [59]. Similarly, deamination of cytosine to form uracil is a common event, which leads to inappropriate pairing with adenine during polymerase extension, thus producing artifactual C:G→T:A mutations, at a frequency approaching 100% [60]. Significantly, DNA damage and the resulting sequencing artifacts occur in strand-specific patterns.

Schmitt et al. recognized that these types of errors could be resolved by exploiting the fact that DNA naturally exists as a double-stranded entity, with one molecule reciprocally encoding the sequence information of its complement. Using this insight and the arising sequencing methodology, termed Duplex Sequencing, Schmitt et al., demonstrated that it is possible to identify and eliminate nearly all sequencing errors by comparing the sequence of individually tagged amplicons derived from one strand of DNA with that of its complementary strand; a base sequenced at a given position is scored only if the read data from each of the two strands match perfectly. The method has a theoretical background error rate of less than one artifactual error per 109 nucleotides and has been used to detect variants at a frequency of 5×10^{-8}.

In principle, Duplex Sequencing can be used with any NGS platform and can call sequence variants when present in an excess of 10 million wild-type sequences [53,56,61]. In contrast, with an error rate of approximately 10^{-2}, the probability of accurately distinguishing a true subclonal variant from a sequencing artifact in an excess of 100 wild-type molecules with NGS is approximately 50%, using standard (Q30)–filtered reads (Figure 1). A real variant at or below these frequencies cannot be resolved by increasing sequencing depth at a single position, as the proportion of errors will not change. Duplex Sequencing, thus, offers an improvement of nearly 5-orders of magnitude over standard Q30-filtered sequencing and 3-orders of magnitude over other tag-based methods. Thus by exploiting the redundant sequence information contained in the complementary strand of a double-stranded DNA molecule, Duplex Sequencing has dramatically increased the precision and power of NGS. Its application will likely improve our understanding of the substructure of biological systems, including human cancers, help to pinpoint mechanisms of mutation generation, modify the catalog of rare variants, dramatically improve our ability to accurately deconvolute complex biological admixtures, and offer the diagnostic accuracy required for the implementation of precision medicine.

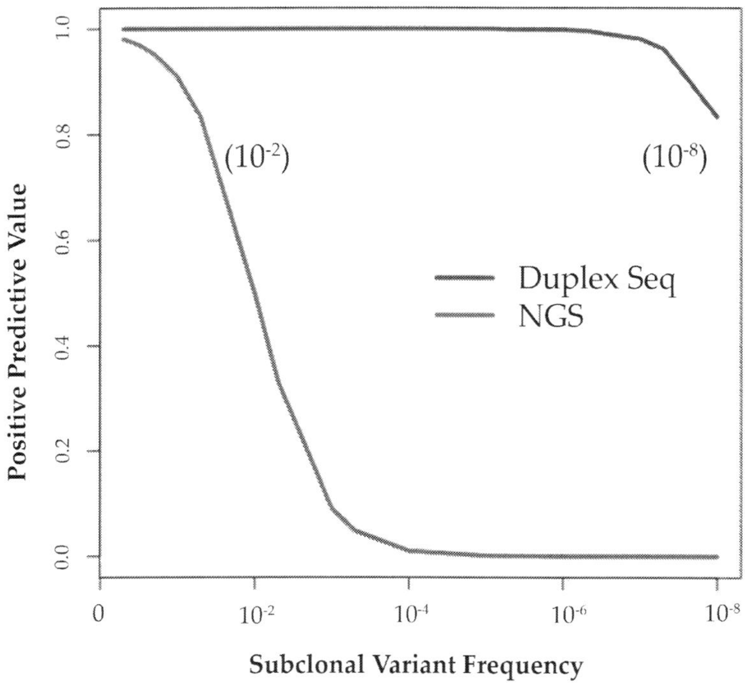

Figure 1: Comparison of the probability that an observed variant is real [54] for subclonal variants using Q30-filtered reads of an Illumina HiSeq2500 (NGS) versus Duplex Sequencing. Error Frequencies of each approach is given in parenthesis. PPV (Positive Predictive Value)=(Expected Number of true positives)/(Expected Total Number of Positive Calls). Note that the PPV is 0.50 for NGS when the variant frequency at a single position is ~1/100, i.e., any variant call has a 50/50 chance of being real hen the frequency of real variants equals the frequency of mistakes invalidity [62].

REFERENCES

1. Loeb LA. Human cancers express mutator phenotypes: origin, consequences and targeting. Nat Rev Cancer. 2011;11:450–457.
2. Shendure J, Ji H. Next-generation DNA sequencing. Nat Biotechnol. 2008;26:1135–1145.
3. Schwartz WB, Wolfe HJ, Pauker SG. Pathology and probabilities: a new approach to interpreting and reporting biopsies. N Engl J Med. 1981;305:917–923.

4. Druley TE, Vallania FL, Wegner DJ, Varley KE, Knowles OL, et al. Quantification of rare allelic variants from pooled genomic DNA. Nat Methods. 2009;6:263–265.
5. LaTuga MS, Ellis JC, Cotton CM, Goldberg RN, Wynn JL, et al. Beyond bacteria: a study of the enteric microbial consortium in extremely low birth weight infants. PLoS One. 2011;6:e27858.
6. Hyman RW, Herndon CN, Jiang H, Palm C, Fukushima M, et al. The dynamics of the vaginal microbiome during infertility therapy with in vitro fertilization-embryo transfer. J Assist Reprod Genet. 2012;29:105–115.
7. Minot S, Sinha R, Chen J, Li H, Keilbaugh SA, et al. The human gut virome: inter-individual variation and dynamic response to diet. Genome Res. 2011;21:1616–1625.
8. Nasu A, Marusawa H, Ueda Y, Nishijima N, Takahashi K, et al. Genetic heterogeneity of hepatitis C virus in association with antiviral therapy determined by ultra-deep sequencing. PLoS One. 2011;6:e24907.
9. Yang J, Yang F, Ren L, Xiong Z, Wu Z, et al. Unbiased parallel detection of viral pathogens in clinical samples by use of a metagenomic approach. J Clin Microbiol. 2011;49:3463–3469.
10. Campbell PJ, Pleasance ED, Stephens PJ, Dicks E, Rance R, et al. Subclonal phylogenetic structures in cancer revealed by ultra-deep sequencing. Proc Natl Acad Sci U S A. 2008;105:13081–13086.
11. Landau DA, Carter SL, Stojanov P, McKenna A, Stevenson K, et al. Evolution and impact of subclonal mutations in chronic lymphocytic leukemia. Cell. 2013;152:714–726.
12. Costello M, Pugh TJ, Fennell TJ, Stewart C, Lichtenstein L, et al. Discovery and characterization of artifactual mutations in deep coverage targeted capture sequencing data due to oxidative DNA damage during sample preparation. Nucleic Acids Res. 2013;41:e67.
13. Ding L, Ley TJ, Larson DE, Miller CA, Koboldt DC, et al. Clonal evolution in relapsed acute myeloid leukaemia revealed by whole-genome sequencing. Nature. 2012;481:506–510.
14. Wang C, Mitsuya Y, Gharizadeh B, Ronaghi M, Shafer RW. Characterization of mutation spectra with ultra-deep pyrosequencing: application to HIV-1 drug resistance. Genome Res. 2007;17:1195–

1201.

15. Carlson CA, Kas A, Kirkwood R, Hays LE, Preston BD, et al. Decoding cell lineage from acquired mutations using arbitrary deep sequencing. Nat Methods. 2011;9:78–80.
16. Ameur A, Stewart JB, Freyer C, Hagström E, Ingman M, et al. Ultra-deep sequencing of mouse mitochondrial DNA: mutational patterns and their origins. PLoS Genet. 2011;7:e1002028.
17. Boyd SD, Marshall EL, Merker JD, Maniar JM, Zhang LN, et al. Measurement and clinical monitoring of human lymphocyte clonality by massively parallel VDJ pyrosequencing. Sci Transl Med. 2009;1:12ra23.
18. Glenn TC. Field guide to next-generation DNA sequencers. Mol Ecol Resour. 2011;11:759–769.
19. Kanagawa T. Bias and artifacts in multitemplate polymerase chain reactions (PCR). J Biosci Bioeng. 2003;96:317–323.
20. Lecroq B, Lejzerowicz F, Bachar D, Christen R, Esling P, et al. Ultra-deep sequencing of foraminiferal microbarcodes unveils hidden richness of early monothalamous lineages in deep-sea sediments. Proc Natl Acad Sci U S A. 2011;108:13177–13182.
21. Mackelprang R, Waldrop MP, DeAngelis KM, David MM, Chavarria KL, et al. Metagenomic analysis of a permafrost microbial community reveals a rapid response to thaw. Nature. 2011;480:368–371.
22. Tillmar AO, Dell'Amico B, Welander J, Holmlund G. A universal method for species identification of mammals utilizing next generation sequencing for the analysis of DNA mixtures. PLoS One. 2013;8:e83761.
23. Schubert M, Ermini L, Der Sarkissian C, Jónsson H, Ginolhac A, et al. Characterization of ancient and modern genomes by SNP detection and phylogenomic and metagenomic analysis using PALEOMIX. Nat Protoc. 2014;9:1056–1082.
24. Out AA, van Minderhout IJ, Goeman JJ, Ariyurek Y, Ossowski S, et al. Deep sequencing to reveal new variants in pooled DNA samples. Hum Mutat. 2009;30:1703–1712.
25. Fan HC, Blumenfeld YJ, Chitkara U, Hudgins L, Quake

SR. Noninvasive diagnosis of fetal aneuploidy by shotgun sequencing DNA from maternal blood. Proc Natl Acad Sci U S A. 2008;105:16266–16271.

26. Chiu RW, Akolekar R, Zheng YW, Leung TY, Sun H, et al. Noninvasive prenatal assessment of trisomy 21 by multiplexed maternal plasma DNA sequencing: large scale validity study. BMJ. 2011;342:c7401.
27. Beck J, Urnovitz HB, Mitchell WM, Schutz E. Next generation sequencing of serum circulating nucleic acids from patients with invasive ductal breast cancer reveals differences to healthy and nonmalignant controls. Mol Cancer Res. 2010;8:335–342.
28. Leary RJ, Kinde I, Diehl F, Schmidt K, Clouser C, et al. Development of personalized tumor biomarkers using massively parallel sequencing. Sci Transl Med. 2010;2:20ra14.
29. Di Mauro E, Saladino R, Trifonov EN. The path to life's origins. Remaining hurdles. J Biomol Struct Dyn. 2014;32:512–522.
30. Frenkel ZM, Trifonov EN. Origin and evolution of genes and genomes. Crucial role of triplet expansions. J Biomol Struct Dyn. 2012;30:201–210.
31. Sobolevsky Y, Guimarães RC, Trifonov EN. Towards functional repertoire of the earliest proteins. J Biomol Struct Dyn. 2013;31:1293–1300.
32. Gerhardt GJ, Takeda AA, Andrighetti T, Sartor IT, Echeverrigaray SL, et al. Triplet entropy analysis of hemagglutinin and neuraminidase sequences measures influenza virus phylodynamics. Gene. 2013;528:277–281.
33. Yang Y, Zhang Y, Jia M, Li C, Meng L. Non-degenerate graphical representation of DNA sequences and its applications to phylogenetic analysis. Comb Chem High Throughput Screen. 2013;16:585–589.
34. Huang T, Zhang J, Xu ZP, Hu LL, Chen L, et al. Deciphering the effects of gene deletion on yeast longevity using network and machine learning approaches. Biochimie. 2012;94:1017–1025.
35. Wu G, Yan S. Prediction of mutations engineered by randomness in H5N1 neuraminidases from influenza A virus. Amino acids. 2008;34:81–90.

36. Kozarewa I, Ning Z, Quail MA, Sanders MJ, Berriman M, et al. Amplification-free Illumina sequencing-library preparation facilitates improved mapping and assembly of (G+C)-biased genomes. Nat Methods. 2009;6:291–295.
37. Vandenbroucke I, Van Marck H, Verhasselt P, Thys K, Mostmans W, et al. Minor variant detection in amplicons using 454 massive parallel pyrosequencing: experiences and considerations for successful applications. Biotechniques. 2011;51:167–177.
38. Vandenbroucke I, Eygen VV, Rondelez E, Vermeiren H, Baelen KV, et al. Minor Variant Detection at Different Template Concentrations in HIV-1 Phenotypic and Genotypic Tropism Testing. Open Virol J. 2008;2:8–14.
39. Flaherty P, Natsoulis G, Muralidharan O, Winters M, Buenrostro J, et al. Ultrasensitive detection of rare mutations using next-generation targeted resequencing. Nucleic Acids Res. 2012;40:e2.
40. Muralidharan O, Natsoulis G, Bell J, Newburger D, Xu H, et al. A cross-sample statistical model for SNP detection in short-read sequencing data. Nucleic Acids Res. 2012;40:e5.
41. Zagordi O, Klein R, Däumer M, Beerenwinkel N. Error correction of next-generation sequencing data and reliable estimation of HIV quasispecies. Nucleic Acids Res. 2010;38:7400–7409.
42. Quail MA, Kozarewa I, Smith F, Scally A, Stephens PJ, et al. A large genome center's improvements to the Illumina sequencing system. Nat Methods. 2008;5:1005–1010.
43. Shen Y, Wan Z, Coarfa C, Drabek R, Chen L, et al. A SNP discovery method to assess variant allele probability from next-generation resequencing data. Genome Res. 2010;20:273–280.
44. Klco JM, Spencer DH, Miller CA, Griffith M, Lamprecht TL, et al. Functional heterogeneity of genetically defined subclones in acute myeloid leukemia. Cancer Cell. 2014;25:379–392.
45. Miner BE, Stöger RJ, Burden AF, Laird CD, Hansen RS. Molecular barcodes detect redundancy and contamination in hairpin-bisulfite PCR. Nucleic Acids Res. 2004;32:e135.
46. McCloskey M, Stöger R, Hansen RS, Laird CD. Encoding PCR products with batch-stamps and barcodes. Biochem Genet. 2007;45:761–767.

47. Casbon JA, Osborne RJ, Brenner S, Lichtenstein CP. A method for counting PCR template molecules with application to next-generation sequencing. Nucleic Acids Res. 2011;39:e81.
48. Jabara CB, Jones CD, Roach J, Anderson JA, Swanstrom R. Accurate sampling and deep sequencing of the HIV-1 protease gene using a Primer ID. Proc Natl Acad Sci U S A. 2011;108:20166–20171.
49. Kivioja T, Vähärautio A, Karlsson K, Bonke M, Enge M, et al. Counting absolute numbers of molecules using unique molecular identifiers. Nat Methods. 2011;9:72–74.
50. Kinde I, Wu J, Papadopoulos N, Kinzler KW, Vogelstein B. Detection and quantification of rare mutations with massively parallel sequencing. Proc Natl Acad Sci U S A. 2011;108:9530–9535.
51. Shiroguchi K, Jia TZ, Sims PA, Xie XS. Digital RNA sequencing minimizes sequence-dependent bias and amplification noise with optimized single-molecule barcodes. Proc Natl Acad Sci U S A. 2012;109:1347–1352.
52. Fu GK, Hu J, Wang PH, Fodor SP. Counting individual DNA molecules by the stochastic attachment of diverse labels. Proc Natl Acad Sci U S A. 2011;108:9026–9031.
53. Lou DI, Hussmann JA, McBee RM, Acevedo A, Andino R, et al. High-throughput DNA sequencing errors are reduced by orders of magnitude using circle sequencing. Proc Natl Acad Sci U S A. 2013;110:19872–19877.
54. Jackson AL, Loeb LA. The mutation rate and cancer. Genetics. 1998;148:1483–1490.
55. Roach JC, Glusman G, Smit AF, Huff CD, Hubley R, et al. Analysis of genetic inheritance in a family quartet by whole-genome sequencing. Science. 2010;328:636–639.
56. Schmitt MW, Kennedy SR, Salk JJ, Fox EJ, Hiatt JB, et al. Detection of ultra-rare mutations by next-generation sequencing. Proc Natl Acad Sci U S A. 2012;109:14508–14513.
57. Lindahl T, Wood RD. Quality control by DNA repair. Science. 1999;286:1897–1905.
58. Preston BD, Albertson TM, Herr AJ. DNA replication fidelity and

cancer. Semin Cancer Biol. 2010;20:281–293.

59. Shibutani S, Takeshita M, Grollman AP. Insertion of specific bases during DNA synthesis past the oxidation-damaged base 8-oxodG. Nature. 1991;349:431–434.

60. Stiller M, Green RE, Ronan M, Simons JF, Du L, et al. Patterns of nucleotide misincorporations during enzymatic amplification and direct large-scale sequencing of ancient DNA. Proc Natl Acad Sci U S A. 2006;103:13578–13584.

61. Kennedy SR, Salk JJ, Schmitt MW, Loeb LA. Ultra-sensitive sequencing reveals an age-related increase in somatic mitochondrial mutations that are inconsistent with oxidative damage. PLoS Genet. 2013;9:e1003794.

62. Greenberg RS. Medical epidemiology. 4th edn Lange Medical Books/McGraw-Hill; New York: 2005.

CHAPTER 11

FOUR GENERATIONS OF SEQUENCING- IS IT READY FOR THE CLINIC YET?

Srilakshmi Srinivasan and Jyotsna Batra

Australian Prostate Cancer Centre- Queensland, Translational Research Institute, Queensland University of Technology, Woolloongabba, AustraliaSanger

ABSTRACT

Next-generation sequencing techniques have revolutionized over the last decade providing researchers with low cost, high-throughput alternatives compared to the traditional Sanger sequencing methods. These sequencing techniques have rapidly evolved from first-generation to fourth-generation with very broad applications such as unraveling the complexity of the genome, in terms of genetic variations, and having a high impact on the biological field. In this review, we discuss the transition of sequencing from the second-generation to the third- and fourthgenerations, and describe some of their novel biological applications. With the advancement in technology, the earlier challenges of minimal size of the instrument, flexibility of throughput, ease of data analysis and short run times are being addressed. However, the need for prospective analysis and effectiveness to test whether the knowledge of any given new variants identified has an effect on clinical outcome may need improvement.

Keywords

NGS, TGS, Genome sequencing, Amplification, Readlength

ABBREVIATIONS

NGS: Next-Generation Sequencing; SNPs: Single Nucleotide Polymorphisms; CNVs: Copy Number Variants; CLIP-seq: Cross-Linking Immunoprecipitation Sequencing; ICGC: International Genome Consortium; TCGA: The Cancer Gene Atlas; TGS-3G: Third- Generation Sequencing; 4G: Fourth-Generation Sequencing; APS: Adenosine 5' Phosphosulfate; PPi: Pyrophosphate; Mb: Mega bases; GGiga SBS: Sequencing by Synthesis; SNVs: Single Nucleotide Variants; MDA: Multiple Displacement Amplification; SLAF-seq: Specific- Locus Amplified Fragment Sequencing; SMS: Single-Molecule Sequencing; PGM: Personal Genome Machine; RRL: Reduced Representation Library; SMS: Single Molecule Sequencing; SMRTTM: Single Molecule Real-Time; tSMS: True Single Molecule Sequencing; ZMW: Zero-Mode Waveguide; FRET: Fluorescence Resonance Energy Transfer; TEM: Transmission Electron Microscopy; Ms: Milliseconds; ELLIDA: Enzymatic Luminometric Inorganic Pyrophosphate Detection Assay; FDA: Food and Drug Administration; CANCP: Solid Tumor Targeted Cancer Gene Panel by Next-Generation Sequencing; VUS: Variants of Unknown Clinical Significance; IVD: In Vitro diagnostic; NIST: National Institute of Standards and Technology microbial genomes and various Mendelian disorders previously thought to be inaccessible, are now possible through these new highthroughput techniques.

INTRODUCTION

The past decade has experienced advancement in the study of genetics, molecular diagnostics and personal medicine through the discovery and improvement of next-generation sequencing (NGS) techniques. Continuous improvement in NGS technologies imply, that the whole genome can be sequenced faster, easier and with a higher accuracy since the advent of Sanger sequencingin 1977. These new techniques are highly poised to disclose the complexity between genetic variants such as single nucleotide polymorphisms (SNPs), copy number variations (CNVs), structural variants, including gene fusions with diseases. Insights into epigenetics, protein-DNA interactions of

NGS encompasses several technologies utilizing distinct sequencing biochemistry approaches and is mainly emphasized by its ability to simultaneously perform millions of sequencing reactions. The most widely used applications of NGS include targeted sequencing and whole transcriptome sequencing [1,2] while CLIP-seq (Cross-linking immunoprecipitation sequencing) a technique to map RNA binding sites for a protein of interest on a genome-wide scale is less frequently used. NGS enables worldwide collaborative efforts like the International Genome Consortium (ICGC) [3] and the Cancer Genome Atlas (TCGA) projects [4] to catalogue many thousands of cancer genomes for several disease types. Research discoveries from these projects have been published and have been non-trivial in improving our knowledge of disease pathogenesis thereby bridging the molecular pathology and personalized medicine [5]. These discoveries have the potential to benefit patients, clinicians and health-care systems to suggest changes of lifestyle behavior and/or in medical interventions to reduce disease related morbidity and mortality [6].

Although NGS have a wide range of biological outcomes, the costs per sample analyses often limit the use of these techniques. Fortunately, recently developed high-throughput techniques reduce the burden of the costs of sequencing. For example, sequencing costs have massively reduced from $5,292.39/Mb in 2001 to $0.06/Mb by April 2013 [7]. It is estimated that the sequencing costs will further reduce with precipitous dropping per-base cost with advancing techniques.

In this review, we discuss common methods and challenges of NGS sequencing, and also their advancement from second-generation to the next two levels, third-generation (3G) and fourth-generation (4G). These techniques have allowed genomics to move from platforms that require PCR amplification of the template prior to sequencing, to single

DNA molecule sequencing without a prior amplification step as in third-generation sequencing techniques, and to a more refined level of the fourth-generation.

SECOND-GENERATION SEQUENCING

NGS techniques are quite diverse but conceptually similar. The preparation of library includes random shearing of DNA followed by ligation with common adaptors. The method used for determining template DNA sequence can be used to classify NGS techniques as follows:

Cyclic Array Sequencing

This technique comprises the early methods of NGS techniques and includes pyrosequencing that relies on the detection of released pyrophosphates as they are incorporated during amplification. The Roche 454 GS20 sequencer introduced the first NGS platform using this technique. The library DNA's attached to 454-specific adaptors are denatured into single strands and captured by amplification beads followed by emulsion PCR [8]. On a picotiter plate, one of dNTPs (dATP, dCTP, dTTP, dGTP) complementary to the template are incorporated by ATP sulfurylase, luciferase, luciferin, DNA polymerase, and adenosine 5' phosphosulfate (APS) and pyrophosphate (PPi) equal to the number of incorporated nucleotides. The incorporated pyrophosphates can be traced by pyrogramTM or enzymatic luminometric inorganic pyrophosphate detection assay (ELLIDA), which corresponds to the order of correct nucleotides that has been incorporated [9]. The unincorporated nucleotides are degraded by apyrase (Figure 1). The read length initially for Roche 454 was 100-250 bp in 2005, generating 200,000+ reads and 20 Mb output per run. In 2008 the Roche 454 was replaced by the 454 GS FLX Titanium system which is capable of generating 700 megabase (Mb) of sequence in 700 bp reads in a 23 hr. run with an accuracy of 99.9% after filtering [10]. In 2009, Roche combined the GS Junior, a bench top system, into the 454 sequencing system and the output was upgraded to 14 gigabases (G) per run [11]. The specific advantage with these systems is the speed which only takes 10 hours to sequence. However, the technique has the disadvantage of having high reagent costs and high error rates for poly-bases longer than 6 bp (**Table 1**).

Table 1: Summary of high-throughput sequencing methods

Method	Advantages	Disadvantages	Ref	Comments
Second-generation sequencing techniques				
454 sequencing	Generates long read lengths and relatively fast run times of the instrument	Poor interpretation of homopolymers leading to errors	[49]	First introduced NGS technique
Illumina (Solexa) Genome Analyzer	Short read length approach and is the most widely used analyzer	Aberrant incorporation of incorrect dNTPs by polymerases	[50]	Low multiplexing ability
HiSeq 2000 (Illumina, CA, USA)	Requires less sample < 1 μg	75 (35-100) bp read lengths. More false positives	[11]	Addition of fluorescent-labeled nucleotides
ABI SOLiD system	Reduction in error rates relative to Illumina NGS system	Have long run times and need for 2-20 μg DNA	[16]	Driven by DNA ligase than polymerase
Polonator G.007	Decode the base by single-base probe in nonamers	In adequate coverage, false-positive SNP selection rate	[18]	Ligation based sequencer
Ion Torrent Sequencing	First platform to eliminate cost and complexity with 4-color optical detection used by other NGS platforms	High accuracy and short run time	[27]	Non-optical DNA sequencing
SLAF-seq	*De novo* SNP discovery with reduced cost and high accuracy	Needs complex instrument	[20]	Double barcode system ensures simultaneous genotyping of large populations
Third-generation sequencing techniques				
PacBio RS (Pacific Biosciences, CA, USA)	No amplification of template DNA required, real-time monitoring of nucleotide incorporation,	High error rates and low reads	[25]	Generates long-read lengths 800-1000 bp
HeliscopeTM Sequencer	Nonbiased DNA sequence	High NTP incorporation error rates	[51]	Single molecule sequencing
Fourth-generation techniques				

Oxford Nanopore	Fastest sequencer whole-genome scan within 15 min	Not much data available, high cost per Mb	[32]	Expanding technique

The cyclic array sequencing technique also involves fluorescent in situ sequencing by synthesis that can determine the template DNA sequence by detecting the exact nucleotide extended by its tagged fluorescent moiety as the sequencing proceeds. The sample preparation involves denaturing of the DNA libraries with adaptors to single strands, grafted to the flow-cell and bridge amplified to form millions of spatially separated clonal DNA fragments. Illumina (Solexa) Genome Analyzer adopts this technique and is the current dominating NGS technique in the market [12]. The first Solexa Genome Analyzer had an output of 1 G/run in 2006 and has gradually increased to 85 G/run within three years by the GAIIx series. The HiSeq 2000 released after 4 years of the first release of the Genome Analyzer and has an output of 600 G/run with an error rate of <2% in average after filtering. The HiSeq2000 is cheaper than the 454 and SOLiD systems, with an ability to handle thousands of samples simultaneously. MiSeq from Illumina also uses this sequencing by synthesis (SBS) technique that combines cluster generation, synthesis and data analysis in single equipment. It can complete sample preparation and analysis within a single day [11]. Efficient mapping of the short reads generated by these sequencers to the reference genome is challenging using traditional methods like BLAST [13] and BLAT [14]. Therefore, recent alignment algorithms such as BWA, which enables variant detection, and Bowtie, that can identify indels and are less prone to false-positive single nucleotide variants (SNVs) around indels, have been developed. These two algorithms and additional novel sequencing pipelines have emerged to help researchers in taking full advantage of NGS technologies [15].

Recently Qiagen (Venlo, Netherlands) have released the whole genome amplification technique that can amplify complex genomes with less bias called the REPLI-g kit. REPLI-g uses Multiple Displacement Amplification (MDA) that involves binding of random hexamers and strand displacement synthesis by the enzyme Phi 29 polymerase. The Phi 29 polymerase can perform uniform amplification with minimized mutation rates and is compatible with many sequencing systems such as the Illumina HiSeq, Roche 454 and many other NGS instruments. With the aim of commercial use of NGS in clinics, Qiagen is planning for commercialization of a new pushbutton automated "sample-to-insight workflow" to target clinical research and the diagnostics market in 2014. This new SBS sequencing system, Gene Reader bench-top sequencer integrates the Qia Cube instrument for DNA extraction, target selection, and sample preparation; Qia Cube NGS for library amplification preparation and bead enrichment; and the Gene Reader. The Gene Reader performs primary analysis - base calling, and generates a FASTQ file; secondary analysis to perform read alignment, variant calling using the CLC Bio software; and tertiary analysis – biological and clinical interpretation using the Ingenuity software. This high-grade integration would likely have a high impact on bridging the gaps between the laboratory based research and clinical application of NGS.

Hybridization using Oligonucleotide Probes

This method originated from Jay Shendure and colleagues in 2005 [16], and is utilized by the ABI SOLiD system since 2006. Repeated cycles of hybridization with oligonucleotide probes allow sequencing of libraries on a SOLiD flow cell. These probes harbor a ligation site (1st base), a cleavage site (5th base), and 4 different fluorescent dyes linked to the last base [17]. The fluorescent signal is recorded as the complementary primer is hybridized to the template strand and the fluorescent signal is lost by the cleavage of the last 3 bases of the probe. The sequence of the DNA template can be deduced after 5 rounds of sequencing using ladder primer sets. The SOLiD system can be used for whole genome sequencing, transcriptome research, targeted sequencing, and epigenome sequencing. The latest SOLiD 5500x1 system has an output of 85 bp, and a high accuracy of 99.99% and 30 G/run.

Another sequencer from complete genomics based on the Polonator G.007 is a ligation-based sequencer. The sequencer can decode the base by single-base probe in nanonucleotides (nanomers) [18]. The primary challenge of the sequencing by hybridization technique involves the probe design and the challenge to avoid crosshybridization of the probe to the incorrect target due to repetitive elements. This may lead to inadequate coverage of the genome and false-positive SNP detection rates [19]. Also, the technique is time consuming, expensive and needs high infrastructure for analysis.

Specific-locus amplified fragment Sequencing (SLAF-seq)

SLAF-seq is a low cost but efficient high-throughput sequence based technique and reduces the complexity of high-quality reference genome libraries that are required for other NGS techniques, and uses the strategy of a reduced representation library (RRL) method [20]. As the method does not require a reference genome sequence, it uses barcode multiplexed sequencing for multiple loci simultaneously, and combines locus-specific amplification and high-throughput sequencing for de novo SNP detection. The double barcode system distinguishes individuals in large populations of about 10,000 samples (**Figure 1**). A study testing the efficiency of the SLAF-seq on rice and soybean genomes have observed the genotyping data to be accurate and the density of the genetic map to be high compared to all the genome data available by other methods so far. Thus, SLAF-seq represents a low-cost large-scale genotyping technique with an important role in genetic association studies.

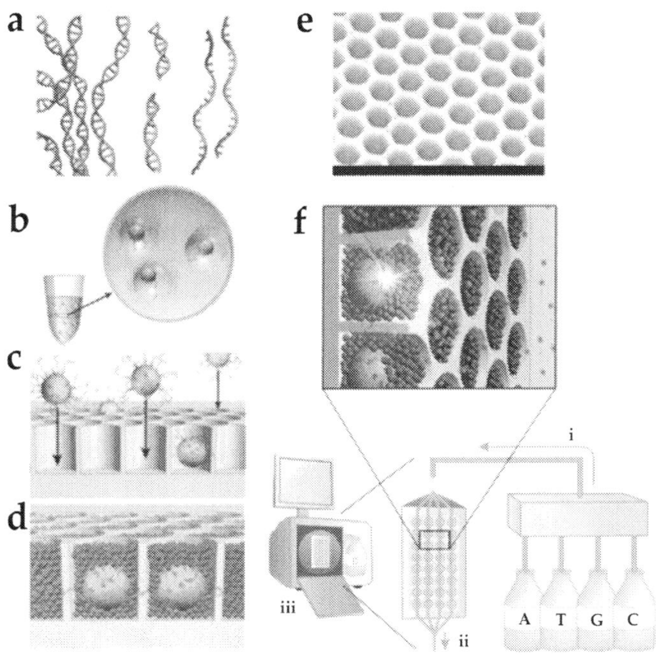

Figure 1: 454 sequencing. (a) Genomic DNA is isolated, fragmented, ligated to adapters and denatured into single strands. (b) DNA fragments are bound to Streptavidin coated magnetic beads under conditions that allow one fragment per bead, the beads are isolated and grouped in the droplets of a PCR-reaction-

mixture in- oil emulsion and PCR amplification occurs within each droplet. (c) The emulsion is broken, the DNA strands are denatured, and beads carrying single-stranded DNA templates are enriched (not shown) and deposited into wells of a fiber-optic slide. (d) Smaller beads carrying enzymes required for a solid phase pyrophosphate sequencing reaction are deposited into each well. (e) Scanning electron micrograph of a portion of a fiber-optic slide, showing fiber-optic cladding and wells before bead deposition. (f) The 454 sequencing instrument consists of the following major subsystems: a fluidic assembly (object i), a flow cell that includes the well containing fiber-optic slide (object ii), a CCD camera-based imaging assembly with its own fiber-optic bundle used to image the fiber-optic slide (part of object iii), and a computer that provides the necessary user interface and instrument control (part of object iii) (Figure reproduced from Rothberg et al with permission from [10]).

Although all the above 2G platforms may be suitable for clinical application, second-generation sequencing machines involve PCR amplification of sheared DNA fragments of about 35-400 bp, which are easily error-prone due to clonal amplification.

THIRD-GENERATION SEQUENCING

Despite the technological differences, the three categories of second-generation sequencing techniques have similar workflows for the sample preparation and analysis [21]. They also harness the need of an amplification step for template DNA as the NGS techniques are not designed to detect single fluorescent events [12]. PCR amplification is associated with PCR bias and has the possibility of base sequence errors or favoring certain sequences over the others. These biases can be avoided if a single molecule is used for sequencing without a prior amplification step. Also, the data generated by NGS techniques is massive, approximately 300 G bases or more by Illumina HiSeq 2000 instrument. The time-to-result is also long (many days) due to the large number of scanning and washing cycles required. Loss of synchronicity with addition of each nucleotide is another disadvantage of the technique and may lead to noise and secondgeneration sequencing errors [22] and short read lengths. The sequencing biochemistry, configuration and generation of array vary for second and third-generation sequencing techniques, few of which are discussed below.

Single-Molecule Sequencing (SMS)

SMS also known as single-template technology provides several advantages over second-generation sequencing. Two devices, Pacific Biosciences Single Molecule Real-time (SMRTTM) sequencing and the Helicos Biosciences true Single Molecule Sequencing (tSMS) were the providers of the first commercial 3G instruments. The techniques utilize the sequencing by synthesis approach, similar to a few of the NGS techniques but differ by not requiring amplification and hence, reduce the sequencing errors due to amplification as in NGS, reducing compositional bias [18] producing long sequences and supporting a short run-around time. The technique uses a DNA polymerase to drive the reaction and is based on real-time imaging of fluorescent labeled nucleotides as they are synthesized along the template DNA molecules [23]. This imaging is performed by dense array of zero mode waveguide (ZMW) nanostructures that allow optical interrogation of single fluorescent molecules (**Figure 2**). Single functioning DNA polymerase is immobilized at the bottom of each ZMW to process fluorescently labeled nucleotide substrates. The four bases (G, C, T, A) are differentially labeled with deoxyribonucleoside pentaphosphate (DN5Ps) substrates. The fluorescent substrate is linked to the phosphate chain rather than the base and therefore the phosphate chain is cleaved when the nucleotide is incorporated into the DNA strand. Thus, on the incorporation of the phospholinked nucleotide, DNA polymerase frees the substrate molecule from the nucleotide when it cleaves the phosphate chain. The label is quickly removed and does not halt the DNA polymerase activity; which halts after incorporating few base-labeled nucleotides. This sequencing technique generated long read lengths of ~1000 bp in 2009. The recent PacBio®RSII sequencer (Pacific Biosciences, California) released in 2013 can generate 8.5 kb reads by combining the P5 DNA polymerase with C3 chemistry (P5-C3), with the longest reads exceeding 30,000 bases. This SMRT sequencer was reported to be least biased and good coverage in extreme GC content (both GC-rich and GC-poor) compared to Illumina and Ion Torrent sequencers [24]. Although the SMRT technique have many advantages over the NGS techniques, a number of challenges still remain such as excess of 5% error rates by insertions and deletions when assembling genomes [25].

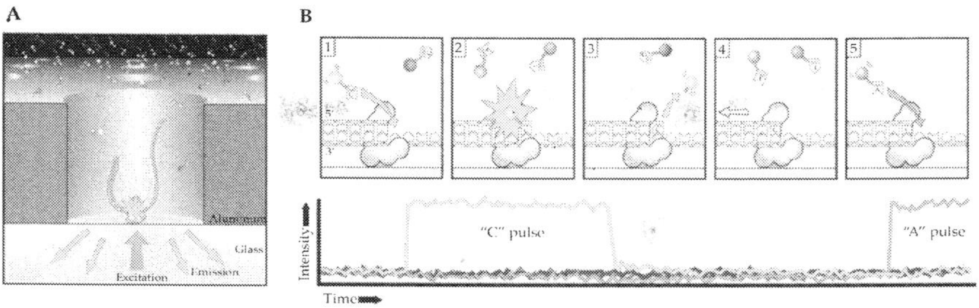

Figure 2: Principle of single molecule-real time sequencing. A) A single molecule of DNA template-DNA polymerase complex is immobilized at the bottom of ZMW illuminated by laser light from the bottom. THE ZMW enables detection of the each incorporated phospholinked nucleotide by the polymerase against the bulk background of nucleotides. B) Schematic representation of the phospholinked dNTP incorporation cycle, with a corresponding expected time trace of detected fluorescence intensity from the ZMW. (1) Cognate association of a phospholinked nucleotide with the template in the polymerase active site, (2) Increased fluorescence output on the corresponding color channel. (3) Formation of a phosphodiester bond liberates the dye-linkerpyrophosphate product that diffuses out of the ZMW, and ends the fluorescence pulse. (4) Translocation of the polymerase to the next position, and (5) binding of the next cognate molecule to the active site, thereby beginning the subsequent pulse (Figure reproduced from Eid et al with permission [24]).

The tSMS technique is slightly different from the SMRT by using a flourophore tagged DNA polymerase which in proximity to a nucleotide, tagged with an acceptor fluorophore, emits a fluorescence resonance energy transfer (FRET) signal. The fluorophore label is released after incorporation. These two 3G techniques, SMRT and tSMS have a series of advantages compared to the second-generation such as higher throughput, longer read lengths to enhance de novo assembly, direct detection of haplotype, whole-chromosome phasing, higher consensus accuracy of identifying rare variants, less sample requirement, making it useful for clinical application. The tSMS still retains many characteristics of second-generation sequencers such as the sequencing approach and chemistry, but the ability to perform direct RNA-sequencing signifies its clear improvement over second-generation techniques.

Transmission electron microscopy (TEM) offered by Halcyon Molecular is another SMS based technique that images and chemically detects atoms comprising DNA templates [26]. Scanning tunneling microscopy can detect DNA bases according to specific electronic differences among the four bases. Although these ideas appear to be straightforward, it has yet to make a long journey due to its challenges to prepare stretched single-stranded DNA on the surface, high cost of microscopes, and its requirement of more specific equipment.

Non-Optical Semi-Conductor Sequencing Technique

The Ion Personal Genome Machine (PGM) launched by Ion torrent, utilizes the power of semiconductor technology to detect protons released as nucleotides are incorporated during synthesis. The system is unique from other techniques by its ability to detect amplification by measuring pH rather than light. The library preparation is similar to 454 sequencing [27]. Sheared DNA strands are linked to adapter sequences; a single DNA template is affixed to a bead (Ion Sphere Particle) and clonally amplified by emulsion PCR. The beads are loaded onto the chip and the dNTPs are flowed over the surface of these beads in a predetermined sequence with zero or more dNTPs ligating during each flow. The 454 sequencing system can introduce 4 nucleotides sequentially, whereas the ion torrent can include 32 nucleotides. This complex flow cycle referred to as Samba improves the synchronicity of clonal templates on the bead at the cost of a flow-sequence not optimized for read length. The protons released for every nucleotide-incorporated decrease the net pH in the surrounding solution that can be measured by an ionic sensor and then converted to a flow value. A base-caller corrects these flow-values for phase and signal loss, normalizes to the key and generates corrected base calls for each flow in each well to produce sequencing reads. Each read is sequentially passed between two signal-based filters to exclude less accurate reads. Per-base quality values are predicted by the Phred method [28] that quantifies the similarity between the phasing model predictions and the observed signal. The Ion PGM is the first system that does not require fluorescence and camera scanning and therefore enables higher output within less time compared to other NGS systems. This technique is also thought to be between the second and third-generation categories [29] because of its wash-and-scan technology. However, its output, speed and use for single-molecule analysis make it a third-generation sequencer.

FOURTH-GENERATION SEQUENCING

The 4G platforms evolved fast, combining single-molecule sequencing of 3G and integrating nanopore technology. Nanopore techniques achieve sequencing without requiring amplification, realtime sequencing without repeated cycles and elimination of synthesis, and so are classified as 4G sequencing techniques.

Nanopore Sequencing

Nanopore sequencing is based on the concept that single DNA molecules can be identified when passing through a tiny nanopore chamber. The nanopores are tiny biopores with nanoscale diameter and can be categorized into biological and solid-state [30].

The biological nanopore is formed by a pore forming protein in the membrane (lipid bilayer) and synthetic materials such as silicon nitride form the solid-state. Oxford Nanopore technology has recently released the commercial nanopore platform that can achieve long-read lengths and have sequenced a phage genome using this technique. The company predicts it can decipher a billion DNA bases in 6 h and sell for US $900 [31]. A single protein nanopore is incorporated in a lipid bilayer across a microwell equipped with electrodes. Sample preparation, detection and analysis are performed in microwells incorporated onto an array chip. The nanopore sequencer has wider applications for protein, RNA and DNA and is mainly adapted for DNA. The sequencing methodologies developed by Oxford include exonuclease and strand sequencing. A cyclodextrin adapter inside the protein nanopore serves as a DNA binding site, and the exonuclease coupled with the nanopore cleaves the individual bases from the DNA strand. The cleaved bases attached to the cyclodextrin are detected (20 ms/base) based on their differences in magnitude of current disruption. Single stranded DNA is passed through the pore in strand sequencing technique and is potentially faster and more accurate than exonuclease sequencing (**Table 1**). The solid-state nanopore is reliably more stable than biological and could be multiplexed to work in parallel on a single device. Many companies are now using the solidstate nanopore technique to perform whole genome sequencing and achieve higher readout within a short time [32]. Complete Genomics have also developed the nanoarray platform where genomic DNA is sonicated to prepare sequencing fragments and adapter sites are inserted followed by circularization of the template (400 bp) and restriction enzyme excision. The circularized templates are amplified by a Phi 29 polymerase and are called a DNA nanoball (DNB) [33]. Up to three billion DNBs are then selectively attached to a silicon chip. This helps in minimizing the reagent costs and increased throughput compared to NGS and earlier TGS instruments. Common probes with standard and extended anchors are hybridized to the DNB array chip and generate read lengths of 5 – 10 bases, resulting in 62 to 70 bases sequences per DNB. Although this method has a higher coverage of the genome compared to Sanger and NGS techniques, the size of the circularized fragments (400 bases), and very short read lengths (6 – 10 bases), prevents complete and accurate genome assembly. In June 2014, Roche announced its intention to purchase Genia, a single molecule tag-based nanopore NGS company. The technique is similar to Oxford Nanopore except that instead of extending the growing amplified DNA product, the Geniananopore allows a cleavable label to enter the pore. The labels induce a change in current through the nanopore, which is read by the instrument. Genia tag-based nanopore technique aims for $100 genomes, and is aiming to release the first platform later this year.

NGS in Clinics

The concept of personalized medicine relies on our access to information on an individual's unique genetic makeup to administer customized therapy. Although the era of NGS has begun, traditional techniques such as PCR and Sanger sequencing are still used for detecting mutations for cancers and other infectious diseases. Few tests such as BRAC Analysis® [34], Melaris® [35] are examples. Initially karyotyping and fluorescent in situ hybridization were used to identify genetic abnormalities, and recently microarrays are successfully used to detect CNVs and SNPs in some Mendelian disorders [36], cancers [37] and common disorders. CNV microarrays suffer the limitation in their ability to assess only large aberrations and therefore cannot identify small genetic abnormalities as observed for many genetic disorders. The SNP microarrays with millions of probes can provide genotype information of common variants as the probes are built only for known information of the genome and cannot assess large aberrations. However, rare variants that may contribute to pathogenesis are often missed. Also, these classical methods do not allow multiplexing of many samples in one lane and targeted or comprehensive coverage of the genome.

The current NGS platforms address the some of the aforementioned limitations of traditional methods, and the clinical community is rapidly embracing NGS techniques to allow treatment based on genetic fingerprints of an individual. NGS techniques use as little as a few picograms of DNA [38] compared to few micrograms for Sanger sequencing [39]. This offers a distinct advantage for use of NGS in the clinic where the amount of available sample is always limiting. Indepth coverage by advanced NGS techniques help in providing information on every nucleotide of the region of interest and therefore novel rare variants of clinical importance can be identified. For example, a rare missense variant not identified by SNP microarrays in the SLC26A3 gene helped in classification of a patient with congenital chloride diarrhea rather than the Bartter syndrome [40]. This example demonstrates the clinical utility of whole-exome sequencing in pinpointing the disease-causing genes in genomic regions identified by GWAS and thus helps in accurate clinical diagnosis.

Further, small bench-top NGS sequencers offer the best use as a clinical diagnostic device as they take less space. The Ion Proton TM, a subsequent generation of Torrent sequencing technology is a benchtop machine for small sample sizes, and can perform whole exome and whole genome sequencing [41]. The current leading platforms include MiSeq from Illumina, and the PGM from Life Technologies which compromise >85% of market share earlier this year. Verinata Health, a non-invasive prenatal testing company leveraged the MiSeq NGS platform to test for fetal chromosomal aneuploidies in high risk pregnancies. This test performed better than standard screening methods for aneuploidies such as Down syndrome, Edwards syndrome, Patau syndrome and other common sex chromosome aneuploidies like Turner syndrome [42]. A bench-top machine from Qiagen, the Gene Reader TM is released this year and is utilized for clinical use and not just for the research setting.

As of late 2013, more than 50 NGS-based diagnostic tests including single-gene assays, multi-gene or multi-transcript panels are in use [43]. Some of the NGS panels employ targeted sequencing methods e.g. include tests available for non-syndromic deafness [44], cancer spinocerebellar ataxias or cardiomyopathies. In April 2014, the Mayo Clinic (http://www.

mayoclinic.org/) launched a new gene panel cancer test called Solid Tumor Targeted Cancer Gene Panel by Next- Generation Sequencing (CANCP) which is a single assay performed that uses formalin-fixed paraffin-embedded tissue to assess common mutations in 50 genes known to affect tumor growth and response to chemotherapy. This test performed using Roche 454 sequencer can be useful for assessing prognosis and guiding treatment of individuals with solid tumors. The data is being used to help determine the clinical trial eligibility for patients with somatic mutations in genes not amenable to current Food and drug administration (FDA)-approved targeted therapies.

Illumina has released the first and only in vitro diagnostic (IVD) NGS platform for accurate and comprehensive cystic fibrosis testing. This system includes two assays: the Cystic Fibrosis 139-Variant Assay that detects 130 clinically relevant CFTR gene variants, and the cystic Fibrosis Clinical Sequencing Assay that accurately captures all variants in the protein coding regions and intron/exon boundaries of CFTR.

One of the major bottlenecks for the adoption of NGS into clinics includes the variability in the performance of variant calling softwares [45,46]. Recently, the National Institute of Standards and Technology (NIST) developed a highly confident variant caller by encouraging the NGS community to share sequencing data of their reference material. This will aid in the standardization of analysis methodologies and better quality checks for assessing false positives and negatives [47,48]. CLC Cancer Research Workbench launched in April 2014, offers cancer-specific, ready-to-use analysis workflows that can be modified and personalized by clinicians and scientists. Ingenuity® Clinical, a new web-based solution is promising to deliver faster, easier-to-use, and high-confidence clinical interpretation of NGS-based tests.

Analysis of variants of unknown clinical significance (VUS) is another major challenge in clinical diagnosis as resources for functional testing on a patient basis is limited. Improving high throughput models of disease using patient-derived samples such as tumor tissues can help to clarify the consequence of the VUS. The commercial success of NGS in the diagnostic era also needs to address other challenges beyond the technology itself, including regulatory approval (such as FDA), ethical concerns, insurance reimbursement, and the ease of analyzing large datasets.

An ethical issue arising from NGS techniques is whether or not to disclose incidental findings. For example, a patient undergoing tests for BRCA1 and 2 gene mutations for breast cancer may show mutations leading to Lynch disease. Whether or not disclosure of this information to that patient is beneficial or harmful has been a dispute. Advancing genomic technologies generate large amounts of sequence data, thus, questioning the right of privacy and confidentiality for an individual. This is because storing large data generated in databases may lead to accidental or unauthorized release of a patient›s unique genomic information. Data sharing across national borders to researchers in other countries also need to be carefully performed as the flaws persist in the current policies as to the handling of biological data.

CONCLUSION

Recently, high-throughput sequencing techniques have progressed vastly, and are in continuous development and improvement. They have a wider application and have a common feature of high throughput data generation. Second-generation sequencing techniques generate massive output but still have limitations of requiring exclusive bioinformatics alignment software for analyses. Further advanced single-molecule sequencing (third-generation) techniques and fourth-generation sequencers have the advantages of accurate and long read lengths without homo-polymer tailing errors, and are moving forward fast to achieve the goal of the $100 genome.

NGS-based diagnostics constitute the major part of personalized medicine and the clinical diagnostic industry. The current target enrichment techniques are more feasible in a clinical diagnostic context. While whole-genome sequencing can identify disease-related genes in clinically well-defined Mendelian diseases (biased for recessive disorders), the application of WGS for heterogenic more complex diseases like cancers need more improvement. Although, the application of NGS with its unique technology is encouraging, several technical and ethical challenges are yet to be addressed. For wider and easier application in clinics, further reduction of the cost of data storage, flexibility in throughput, data analysis, interpretation and reporting are needed. A bench-top sequencer such as the Gene Reader (Qiagen), which can be used with just a push-button option, could be more beneficial in a clinical setting for genomic and molecular diagnostics. Such systems would be easier and more amenable for translating research from "bench to bedside."

We envisage that in 10 years, we may reach a point where analytic validity of sequencing technologies are high, with ease of access to clinical interpretation of genomic data and knowledge about patient responses to genomic information.

REFERENCES

1. MardisER,Wilson RK (2009) Cancer genome sequencing: a review. Hum Mol Genet 18: 163-168.
2. Bras JR, Guerreiro R, Hardy J (2012) Use of next-generation sequencing and other whole-genome strategies to dissect neurological disease. Nat Rev Neurosci13:453-464.
3. Guyer M, Hemsley FM, Jennings JL, Kerr D, Klatt P, et.al (2010) International network of cancer genome projects. Nature 464 : 993-998
4. www.http://cancergenome.nih.gov/
5. Harris TJ, McCormick F (2010) The molecular pathology of cancer. Nat Rev ClinOncol7:251-265.
6. Mwenifumbo JC, Marra MA (2013) Cancer genome-sequencing study design. Nat Rev Genet 14: 321-332.
7. SpurdleAB,Lakhani SR, Da Silva LM, Balleine RL, kConFab Investigators, et.al (2010) Bayes analysis provides evidence of pathogenicity for the BRCA1 c.135-1G>T (IVS3-1) and BRCA2 c.7977-1G>C (IVS17-1) variants displaying in vitro splicing results of equivocal clinical significance. Hum Mutat 31: 1141-1145.
8. Dressman D, Yan H, Traverso G, Kinzler KW, Vogelstein B, et.al (2003) Transforming single DNA molecules into fluorescent magnetic particles for detection and enumeration of genetic variations.ProcNatlAcadSci USA 100: 8817-8822.
9. Ronaghi M, Karamohamed S, Pettersson B, Uhlen M, Nyren P (1996) Real-time DNA sequencing using detection of pyrophosphate release. Analytical biochemistry242: 84-89.
10. Margulies M, Egholm M, Altman WE, Attiya S, Bader JS, et

al (2005) Genome sequencing in microfabricated high-density picolitre reactors. Nature437: 376-380

11. Liu L, Li Y, Li S, Hu N, He Y, et.al (2012) Comparison of next-generation sequencing systems. J Biomed Biotechnol 2012:251364
12. Metzker ML (2010) Sequencing technologies - the next generation. Nat Rev Genet 11:31-46.
13. Altschul SF, Gish W, Miller W, Myers EW, Lipman DJ (1990) Basic local alignment search tool. Journal of molecular biology215: 403-410.
14. Kent WJ (2002) BLAT--the BLAST-like alignment tool. Genome research 12: 656-664.
15. Cox DG, Simard J, Sinnett D, Hamdi Y, Soucy P et.al (2011) Common variants of the BRCA1 wild-type allele modify the risk of breast cancer in BRCA1 mutation carriers. Hum Mol Genet20: 4732-4747.
16. Shendure J, Porreca GJ, Reppas NB, Lin X, McCutcheon JP, et al (2005) Accurate multiplex polony sequencing of an evolved bacterial genome. Science309:1728-1732.
17. Mardis ER (2008) The impact of next-generation sequencing technology on genetics. Trends Genet24: 133-141.
18. Timp W, Mirsaidov UM, Wang D, Comer J, Aksimentiev A, et al (2010) Nanopore Sequencing: Electrical Measurements of the Code of Life. IEEE Trans Nanotechnol9: 281-294.
19. Patil N, Berno AJ, Hinds DA, Barrett WA, Doshi JM, et.al (2001) Blocks of limited haplotype diversity revealed by high-resolution scanning of human chromosome 21. Science294:1719-1723.
20. Sun X, Liu D, Zhang X, Li W, Liu H, et al (2013) SLAF-seq: an efficient method of large-scale de novo SNP discovery and genotyping using high-throughput sequencing.PLoS One, 8: 58700
21. Shendure J, Ji H (2008) Next-generation DNA sequencing. Nat Biotechnol26: 1135-1145.
22. Whiteford N, Skelly T, Curtis C, Ritchie ME, Löhr A, et.al (2009) Swift: primary data analysis for the IlluminaSolexa sequencing platform. Bioinformatics 25: 2194-2199.
23. Eid J, Fehr A, Gray J, Gray J, Lyle J, et al (2009) Real-time DNA

sequencing from single polymerase molecules. Science323:133-138
24. Ross MG, Russ C, Costello M, Hollinger A, Lennon NJ,et. al (2013) Characterizing and measuring bias in sequence data. Genome Biol14: R51.
25. Roberts RJ,Carneiro MO, Schatz MC (2013) The advantages of SMRT sequencing. Genome Biol14: 405.
26. Krivanek OL, Chisholm MF, Nicolosi V, Pennycook TJ, Corbin GJ,et.al (2010) Atom-by-atom structural and chemical analysis by annular dark-field electron microscopy. Nature464: 571-574.
27. Rothberg JM, Hinz W, Rearick TM, Schultz J, Mileski W, et. al (2011) An integrated semiconductor device enabling non-optical genome sequencing. Nature475: 348-352.
28. Ewing B, Green P (1998) Base-calling of automated sequencer traces usingphred. II. Error probabilities Genome Res 8:186-194.
29. Barzon L, Lavezzo E, Militello V, Toppo S, Palù G (2011) Applications of next-generation sequencing technologies to diagnostic virology. Int J MolSci12: 7861-7884
30. Niedringhaus TP, Milanova D, Kerby MB, Snyder MP, Barron AE (2011) Landscape of next-generation sequencing technologies. Anal Chem83: 4327-4341.
31. Pennisi E (2012) Genome sequencing. Going solid-state Science336: 536.
32. Thomas CS, Glassman MJ, Olsen (2011) BDSolid-state nanostructured materials from self-assembly of a globular protein-polymer diblock copolymer. ACS Nano5: 5697-5707.
33. Anderson JP, Reynolds BL, Baum K, Williams JG (2010) Fluorescent structural DNA nanoballs functionalized with phosphate-linked nucleotide triphosphates. Nano Lett10: 788-792
34. Frank TS, Manley SA, Olopade OI, Cummings S, Garber JE, et.al (1998) Sequence analysis of BRCA1 and BRCA2: correlation of mutations with family history and ovarian cancer risk. J ClinOncol, 16: 2417-2425.
35. Glendening JM, Flores JF, Walker GJ, Stone S, Albino AP, et.al (1995) Homozygous loss of the p15INK4B gene (and not the p16INK4 gene) during tumor progression in a sporadic melanoma

patient. Cancer Res55: 5531-5535.

36. Park JJ, Oh BR, Kim JY, Park JA, Kim C,et. al (2011) Copy number variation of beta-defensin genes in Behcet's disease. ClinExpRheumatol29:S 20-23.
37. PanditaA, Bayani J, Paderova J, Marrano P, Graham C, et.al (2011) Integrated cytogenetic and high-resolution array CGH analysis of genomic alterations associated with MYCN amplification.Cytogenet Genome Res 134: 27-39.
38. Parkinson NJ,Maslau S, Ferneyhough B, Zhang G, Gregory L,et al (2012) Preparation of high-quality next-generation sequencing libraries from picogram quantities of target DNA. Genome Res22: 125-133
39. Lander ES, Linton LM, Birren B, Nusbaum C, Zody MC, et al (2001) Initial sequencing and analysis of the human genome. Nature409: 860-921.
40. Choi M, Scholl UI, Ji W, Liu T, Tikhonova IR, et al (2009) Genetic diagnosis by whole exome capture and massively parallel DNA sequencing.ProcNatlAcadSci USA106: 19096-19101
41. Merriman B, Rothberg JM et al (2012) Progress in ion torrent semiconductor chip based sequencing. Electrophoresis 33: 3397-3417
42. Bianchi DW, Parker RL, Wentworth J, Madankumar R, Saffer C, et.al (2014) DNA sequencing versus standard prenatal aneuploidy screening. N Engl J Med370: 799-808.
43. Rehm HL (2013) Disease-targeted sequencing: a cornerstone in the clinic. Nat Rev Genet14: 295-300.
44. Shearer AE, DeLuca AP, Hildebrand MS, Taylor KR, Gurrola J, et al (2010) Comprehensive genetic testing for hereditary hearing loss using massively parallel sequencing.ProcNatlAcadSci USA 107: 21104-21109
45. Gullapalli RR, Desai KV, Santana-Santos L, Kant JA, Becich MJ (2012) Next generation sequencing in clinical medicine: Challenges and lessons for pathology and biomedical informatics. J Pathol Inform3:40.
46. Gullapalli RR, Lyons-Weiler M, Petrosko P, Dhir R, BecichMJ,et.

al (2012) Clinical integration of next-generation sequencing technology. Clin Lab Med32: 585-599.

47. Zook JM, Samarov D, McDaniel J, Sen SK, Salit M (2012) Synthetic spike-in standards improve run-specific systematic error analysis for DNA and RNA sequencing. PLoS One7: e41356
48. Zook JM, Chapman B, Wang J, Mittelman D, Hofmann O, et.al (2014) Integrating human sequence data sets provides a resource of benchmark SNP and indel genotype calls. Nat Biotechnol32: 246-251
49. Wang J, Lin M, Crenshaw A, Hutchinson A, Yeager M, et.al (2009) High-throughput single nucleotide polymorphism genotyping using nanofluidic Dynamic Arrays. BMC genomics 10: 561
50. Peters EJ, McLeod HL (2008) Ability of whole-genome SNP arrays to capture 'must have' pharmacogenomic variants. Pharmacogenomics9: 1573-1577.
51. Harris TD, Buzby PR, Babcock H, Beer E, Bowers J,et.al (2008) Single-molecule DNA sequencing of a viral genome. Science320:106-109.

CHAPTER 12

NEXT GENERATION SEQUENCING TECHNOLOGIES IN MALARIA

Vathsala Palakkod Govindan and Krishna Murthy P

Indian Institute of Science, Bangalore, Karnataka, India

ABSTRACT

The research outlook in life sciences is futuristic, thanks to next generation sequencing technologies. With regard to malaria research, use of next generation sequencing techniques, a tool of great value for study of various infectious diseases, have help find multidrug resistance cause of anti-malarial drugs. Earlier efforts toward sequencing of genes were laborious, expensive and time consuming. At present, advancement in the field of next generation sequencing technologies are changing the horizon and promising an era of elevated human health. In this mini review, the present and potential applications of next generation sequencing technologies in malaria research are discussed.

Keywords

Next generation sequencing, Malaria, *Plasmodium*, Drug resistance, Human health, Infectious diseases

INTRODUCTION

Malaria, an illness caused by protozoan parasites (Plasmodium) and transmitted by mosquitoes, is the main cause of death in children and pregnant women [1]. Most affected areas are underdeveloped countries with 60% mortality rate [2,3]. However, with several advancements in control and management of the disease, the deaths in children, worldwide, below the age of five have fallen by 47% to 54% since 2000 [4]. The species of parasites that causes malaria include *P. falciparum, P. vivax, P. ovale* and*P. malariae* . Of these, *P. falciparum* is widespread in affected regions and causes most deaths. In this regard, new technologies and ideas are explored in search of new treatments and vaccines against malaria. One such approach is whole genome sequencing and analysis of malaria parasites. As reported by Gardner et al. [5], sequence analysis of *P. falciparum* showed the parasite consists of a 23-megabase genome with 14 chromosomes, encoding approximately 5,300 genes with unusual complexity compared to other eukaryotic genomes. Two key unusual features of *P. falciparum* genome are AT-rich content (82%) and poor correlation between the length of a gene and the functional domains of gene due to lack of intron splicing [6]. Owing to composition of genomes, polymorphisms and divergence in size, both within and different species of malaria, the current information of *P. falciparum* genome limits us from extrapolating to other species of malaria. Nevertheless, the rapid pace in developments of next generation sequencing technologies could help to solve this problem.

The most favored and widely used approaches for sequencing are Maxam and Gilbert chemical degradation and Sanger dideoxynucleotide methods however, the techniques are not without constraints [7,8]. The biggest limitations are in terms of highthroughput, large-scale, efficient gene sequencing. With the growing technological advancements, techniques such as large-scale DNA sequencing without gels and reversible terminators were developed with focus on speed and efficiency [9]. On the same scale, platforms such as Illumina/Solexa, ABI/SOLiD,

454/Rochie and Helicos are in the forefront in providing unmatched possibilities for high-throughput genomic research. By generating five million base sequences in hours, 'Pyrosequencing' developed at 454 Life Sciences is one such example that shows the enormous advantage the platform provides for gene sequencing [10].

NEXT GENERATION SEQUENCING APPLICATIONS IN MALARIA

Next generation sequencing (NGS) may overrule the Sanger method of sequencing and expected to advance and make the drug discovery process more rapid. The impact of next generation sequencing techniques in dealing with complex biological problems is huge. One such impact is seen in identification of the sequence changers in drug resistant malaria, HIV, Tuberculosis and cancer. Individual loci associated with drug resistance to many anti-malarials have been identified but their mechanisms have not been elucidated. In addition, NGS technology will create tremendous impact in medical sciences. Understanding of multiple non-genetic and genetic mechanisms of drug resistance would help us in efficient designing and pairing of future anti-malarials for clinical use.

Drug resistance remains a major public health challenge for malaria treatment and eradication. With the completion of malaria parasite genome sequencing, it has leads to greater understanding of the complex biological processes in malaria parasites. Recent technologies with NGS considers a "High-Throughput Technology" which is cost and exponentially effective with first, second generation HT-NGS and third generation sequencing [11-14]. At present, research is being increasingly focused on identifying multidrug resistant cause of antimalarial drugs by employing next generation sequencing techniques, which is of great value for infectious diseases such as malaria.

Whole genome shotgun sequencing of a clone (3D7) of *P. falciparum* has generated a wealth of information about its genome structure, proteome and metabolic potential of several predicted proteins [5]. Likewise, a series of initiatives have begun to generate data on other species of malaria [15], thereby, opening a path for drug resistance and gene discovery despite the huge divergence shown between the species. In addition, a single-cell genomics approach has been established for sequencing of malaria parasite genome sequences directly from patients' blood, which led us to identify and understand drug resistance haplotypes from multiple genotype infections [16]. The variation in haplotypes can also been used to estimate the origin of parasites [17]. On the other hand, genome sequence data from *P. falciparum* have led to comparative studies in understanding characteristic biased amino acid composition of proteins of *P. falciparum* with other eukaryotes [18], which in turn could help understand the specificity of the disease and development of new drug targets.

Methodological advances in NGS such as sequencing by synthesis (SBS) [19], Polony sequencing [20], Nanopore sequencing, sequencing by hybridization (SBH) and sequencing by ligation (SOLiD) [21,22] in conjunction with advanced bioinformatics tools have been contributory to solving fundamental puzzles of drug discovery. NGS technologies in PiggyBac transposon system have helped the system to become a powerful tool in studying gene functions in malaria parasite [23]. The imputation of whole exome sequencing along with lower frequency analysis of variants in the population gives an opportunity to identify rare coding variations in complex traits of several diseases [24]. The new tools of gene expression of Plasmodium proteins measured directly from the tissue during human infection and metagenomic application of NGS by new culture independent detection of pathogens have lead to better understanding of the hostparasite interactions [25]; thus a powerful future application of NGS to combat malaria infections and restore health.

CONCLUSION

Enhanced platforms of NGS are the way forward for further understanding of malaria host-parasite interactions, vector biology and future prospects for the global elimination of malaria. Continued advances in technologies such as single-molecule sequencing and powerful bioinformatics tools to resolve mixed infections and the accurate, rapid response platform provided by NGS will become cornerstones of malaria infection control. In near future, we envision of clinical laboratory setups furnished with high performance genome sequencing platforms that would enable us to derive *Plasmodium* genomes by direct sequencing of clinical samples with automation to identify drug resistant malaria species. We need to look no further than the current challenges of drug resistance; malaria will never be a solved problem because of mimic played by the parasite rather only by substantial effect offered by NGS with the disease.

REFERENCES

1. Aditya NP, Vathsala PG, Vieira V, Murthy RSR, Souto EB (2013) Advances in nanomedicines for malaria treatment. Adv Colloid Interface Sci 201-202: 1-17.
2. Korenromp EL, Williams BG, Gouws E, Dye C, Snow RW (2003) Measurement of trends in childhood malaria mortality in Africa: An assessment of progress toward targets based on verbal autopsy. Lancet Infect Dis 3: 349-358.
3. Snow RW, Guerra CA, Noor AM, Myint HY, Hay SI (2005) The global distribution of clinical episodes of Plasmodium falciparum malaria. Nature 434: 214-217.
4. Organization WH (2014) World Malaria Report 2014.
5. Gardner MJ, Hall N, Fung E, White O, Berriman M, et al. (2002) Genome sequence of the human malaria parasite Plasmodium falciparum. Nature 419: 498-511.
6. Forsdyke DR (2016) Evolutionary Bioinformatics. (3rdedn.) Springer.
7. Maxam AM, Gilbert W (1977) A new method for sequencing DNA. Proc Natl Acad Sci USA 74: 560-564.
8. Sanger F, Nicklen S, Coulson AR (1977) DNA sequencing with chain-terminating inhibitors. Proc Natl Acad Sci U S A 74: 5463-5467.
9. Ansorge WJ (1991) Next-generation DNA sequencing techniques. New Biotechnology 25: 195-203.
10. Margulies M, Egholm M, Altman WE, Attiya S, Bader JS, et al. (2005) Genome sequencing in microfabricated high-density picolitre reactors. Nature 437: 376-380.
11. Le Roch KG, Zhou Y, Blair PL, Grainger M, Moch JK, et al. (2003) Discovery of gene function by expression profiling of the malaria parasite life cycle. Science 301: 1503-1508.
12. Carlton JM, Adams JH, Silva JC, Bidwell SL, Lorenzi H, et al.

(2008) Comparative genomics of the neglected human malaria parasite Plasmodium vivax. Nature 455: 757-763.

13. Dharia NV, Bright AT, Westenberger SJ, Barnes SW, Batalov S, et al. (2010) Whole-genome sequencing and microarray analysis of ex vivo Plasmodium vivax reveal selective pressure on putative drug resistance genes. Proc Natl Acad Sci USA 107: 20045-20050.
14. Khan SM, Franke-Fayard B, Mair GR, Lasonder E, Janse CJ, et al. (2005) Proteome analysis of separated male and female gametocytes reveals novel sex-specific Plasmodium biology. Cell 121: 675-687.
15. Carlton JM, Carucci DJ (2002) Rodent models of malaria in the genomics era. Trends Parasitol 18: 100-102.
16. Nair S, Nkhoma SC, Serre D, Zimmerman PA, Gorena K, et al. (2014) Single-cell genomics for dissection of complex malaria infections. Genome Res 24: 1028-1038.
17. Anderson TJC, Roper C (2005) The origins and spread of antimalarial drug resistance: lessons for policy makers. Acta Trop 94: 269-280.
18. Pizzi E, Frontali C (2001) Low-complexity regions in Plasmodium falciparum proteins. Genome Res 11: 218-229.
19. Shendure J, Ji H (2008) Next-generation DNA sequencing. Nat Biotechnol 26: 1135-1145.
20. Mitra RD, Church GM (1999) In situ localized amplification and contact replication of many individual DNA molecules. Nucleic Acids Res 27: e34.
21. Drmanac R, Drmanac S, Chui G, Diaz R, Hou A, et al. (2002) Sequencing by hybridization (SBH): advantages, achievements, and opportunities. Adv Biochem Eng Biotechnol 77: 75-101.
22. Lagerqvist J, Zwolak M, Di Ventra M (2006) Fast DNA sequencing via transverse electronic transport. Nano Lett 6: 779-782.
23. Cao Y, Rui B, Wellems DL, Li M, Chen B, et al. (2013) Identification of piggyBac-mediated insertions in Plasmodium berghei by next generation sequencing. Malar J 12: 287.
24. Auer PL, Johnsen JM, Johnson AD, Logsdon BA, Lange LA, et al. (2012) Imputation of exome sequence variants into population-based samples and blood-cell-trait-associated loci in African Americans: NHLBI GO Exome Sequencing Project. Am J Hum

Genet 91: 794-808.

25. Van Tyne D, Tan Y, Daily JP, Kamiza S, Seydel K, et al. (2014) Plasmodium falciparum gene expression measured directly from tissue during human infection. Genome Med 6: 110.

CHAPTER 13

THE ROLE OF OPTIMAL DETECTION OF CNAS AND ERROR ANALYSIS USING NEXT GENERATION SEQUENCING

Jorge Munoz-Minjares and Yuriy S Shmaliy

Department of Electronics Engineering, Guanajuato University, Salamanca, Mexico

SHORT COMMUNICATION

Next generation sequencing is a concept that involves several different modern technologies to sequence large stretches of deoxyribonucleic acid (DNA) with wide advantage in throughput and scale compared to antiques sequencing technologies.

Main benefits of the next generation sequencing (NGS) are recognized as follows [1]:

- High-resolution, which allows obtaining a base-by-base view of a gene, exome, or genome.
- Quantitative measurements based on signal intensity.
- Detection all types of genomic alterations, including single nucleotide variants, insertions and deletions, copy number changes,

and chromosomal aberrations.

- High throughput and flexibility in scaling and sequencing multiple samples simultaneously.

These characteristics have brought revolutionary advances to genetic field and resulted in the development of a wide variety of methods, which allowed researchers to ask virtually any question related to the genome, transcriptome, or epigenome of any organism. But in spite of advantages of the NGS technology, an important flaw inherent to earlier developed methods still remains. The NGS data are contaminated by intensive noise that requires using efficient methods of statistical signal processing and bioinformatics to eliminate undesirable information, so that NGS data could finally be used by researchers to make a clinical interpretation.

The methods, which are developed to study the heritable or acquired alterations in the DNA are called genomics. Some examples of this technology are the whole-genome Sequencing, exome sequencing, de novo sequencing and targeted sequencing. The whole genome association studies aim to identify the genetic basis of traits and disease susceptibilities using SNP microarrays that capture most of the common genetic variation in the human population [2]. In [3], one can find a report related to the genome-wide copy number variations, single nucleotide mutations and DNA methylation findings that might be specific in African-American colorectal cancer patients.

Single-nucleotide polymorphism (SNP) arrays simultaneously define the copy-number changes and allelic imbalances (including LOH) occurring in a tumor, at high resolution and throughout the whole genome [4]. SNP arrays are presently one of the most efficient technologies for the identification of copy number alterations [5].

Accurate detection of copy number alterations-aberrations (CNAs) with high accuracy is one of the main objectives of the technologies based on the NGS. The CNAs represent changes in the chromosomal structure, which result in gains or losses in copies of sections of DNA and are often associated with many types of cancer [6]. What many issues face in practice of the CNA detection is that an intensive noise does not allow for an accurate detection of the breakpoints and precise estimation of the segmental levels. In fact, accuracy in the estimates of the segmental levels is strongly affected by the segmental noise [7], which is widely recognized to be white Gaussian [8]. In view of the a fore mentioned issues, many approaches have been developed during decades to provide efficient NGS data denoising while preserving edges in the CNA structures.

Our principal research is based on the study of the statistical characteristics of data provided using the NGS technology and on the development of the confidence probabilistic masks, which allow improving the CNA estimates provided by diverse estimators. Specifically, a great attention is paid to the probabilistic analysis of jitter in the breakpoints of the CNA profiles [9] referring to the following widely recognized properties of the CNA function [8]:.

- It is piecewise constant (PWC) and sparse with a small number of alterations on a long base-pair length.
- Its constant values are integer, although this property is not survived in the log R Ratio.
- The measurement noise in the log R Ratio is highly intensive and can be modelled as additive white Gaussian.

Based on our recent studies of the confidence limits for the stepwise signals measured in white Gaussian noise, we have developed an efficient algorithm for computing the confidence upper and lower boundary masks to guarantee an existence of genomic changes in certain regions with required probability. The masks are formed based on the segmental noise Gaussian distribution and jitter distribution in the breakpoints. We have shown that, in the first order approximation, the jitter in the breakpoints is distributed with the discrete skew Laplace law, which gives satisfactory results for the signal-to-noise ratio (SNR) exceeding unity. Because lower SNR levels are also of high importance, we have proposed several approximations for the jitter distribution with low and highly low SNR levels. The approximations we have developed are based on the modified Bessel function of the second kind and zeroth order. We have also modified the discrete skew Laplace law to have the noise variance dependent on the displacement with respect to the candidate breakpoint. We suggest combining these masks with estimates in order to give medical experts more information about the true CNA structures. Extensive investigations of the NGS-based measurement data provided using the confidence masks ensure that there is always a probability that some changes detected by an estimator do not exist and can be ignored

[10], while some others exist in certain regions and not at exact points predicted by an estimator [11].

In conclusion, the modern NGS-based technologies have provided a high resolution in sequencing that, however, caused an essential increase in the computational complexity of data analysis. On the other hand, while processing data obtained by the NGS, an estimator discovers more subtle chromosomal effects than before that cause another problem. In the presence of intensive noise, not all of the detected CNAs may exist with a probability required. Moreover, many detected segments and breakpoints may exist in certain regions specified by the confidence probability rather than at exact points suggested by an estimator. For this reason, methods of statistical signal processing and bioinformatics will play a more essential role with further development of the NGS-based technologies. This relates to optimal and robust algorithms supplied with methods of error analysis. Final results produced by these techniques must be tested and interpreted by experienced genetic experts.

REFERENCES

1. Illumina (2014) An introduction to next-generation sequencing technology.
2. Pauline C Ng, Kirkness EF (2010) Whole Genome Sequencing. Genetic Variation, Volume 628 of the series methods in molecular biology, Chapter 12. pp. 215-226.
3. Brim H, Ashktorab H (2016) Genomics of colorectal cancer in African Americans. Next Generat Sequenc & Applic 3: 133.
4. Dutt A, Beroukhim R (2017) Single nucleotide polymorphism array analysis of cancer. Curr Opin Oncol 19: 43-49.
5. Engle LJ, Simpson CL, Landers JE (2006) Using high-throughput SNP technologies to study cancer. Oncogene 25: 1594-1601.
6. Redon R, Ishikawa S, Fitch KR, Feuk L, Perry GH, et al. (2006) Global variation in copy number in the human genome. Nature 444(7118): 444–454.
7. Munoz-Minjares JU, Shmaliy YS, Cabal AJ (2014) Noise studies in measurements and estimates of stepwise changes in genome DNA chromosomal structures. Adv Appl Pure Math 212-221.

8. Pique-Regi R, Ortega A, Tewfik A, Asgharzadeh S (2012) Detection changes in the DNA copy number. IEEE Signal Processing Mgn 29: 98-107.
9. Munoz-Minjares JU, Shmaliy YS (2013) Approximate jitter probability in the breakpoints of genome copy number variations. Electrical Eng computing science and autom control (CCE), Mexico City, Mexico 128-131.
10. Munoz-Minjares JU, Cabal-Aragon J, Shmaliy YS (2014) Confidence masks for genome DNA copy number variations in applications to HR–CGH array measurements. Biomed Signal Process Contr 13: 337-344.
11. Munoz-Minjares JU, Shmaliy YS (2017) Improving estimates of the breakpoints in genome copy number alteration profiles with confidence masks. Biomed Signal Process Contr 10: 238-248.

CHAPTER 14

BRINGING NEXT-GENERATION SEQUENCING ONCOLOGY TESTS INTO THE DIAGNOSTIC SETTING

Chee-Seng Ku

Department of Medical Epidemiology and Biostatistics, Karolinska Institutet, Stockholm, Sweden

INTRODUCTION

Since the completion of the Human Genome Project (HGP) in 2003 and the appearance of the first next-generation sequencer (i.e. Roche 454 GS20) on the market [1], it has been widely anticipated that the rate of genomic discovery would increase exponentially. Genomic discovery in the context of human genetics encompasses the identification of new variants or mutations underlying human disease, from rare Mendelian disorders to common complex diseases including cancer, and pharmacogenomics for drug responses [2-4]. It has been envisaged that these 'disruptive technologies' will eventually be translated into the diagnostic setting. Thus, the term 'genomic medicine' has been widely used to describe this rapidly evolving field i.e. applying genomic information in the context of furthering the clinical care of patients. Pharmacogenomics is that branch of genomic medicine that specializes in applying genomic information in the context of therapeutic decision-making [5,6].

The completion of the HGP and the advent of multiple Next-Generation Sequencing (NGS) technologies in its wake have spawned a significant number of new developments in genomic medicine. One of the first examples was provided by the sequencing of an entire cancer genome by means of NGS to determine the genetic aberration in a patient with a diagnosis of Acute Myeloid Leukaemia (AML) of unclear subtype. This Whole-Genome Sequencing (WGS) analysis detected a novel insertional translocation on chromosome 17 which generated a pathogenic PML-RARA gene fusion, thereby confirming a diagnosis of acute promyelocytic leukemia (a subtype of AML with a favorable prognosis), and making possible the improved management of the patient [7]. The combination of NGS technologies with new sequence enrichment methods has also made Whole-Exome Sequencing (WES) feasible. The impact of WES on clinical management was first evidenced by the case of a patient, diagnosed with an intractable inflammatory bowel disease through the identification of a missense mutation in the XIAP gene, in whom allogeneic hematopoietic progenitor cell transplantation was subsequently performed [8].

In addition to WGS and WES, rapid advances have also been made in the application of NGS technologies in targeted sequencing of a panel of genes to detect somatic mutations in tumor tissue, which is important for therapeutic decision-making in the context of molecular targeted drugs. For example, somatic mutations in EGFR inform the prescription of drugs (such as erlotinib and gefitinib) targeting the gene product, a tyrosine kinase [9,10]. However, detecting rare (yet clinically actionable) somatic mutations has been very challenging using conventional methods such as Sanger sequencing. This is due to a combination of tissue heterogeneity and genetic heterogeneity, as well as temporal heterogeneity i.e. tumor evolution; taken together, these account for somatic mutations often being characterized by low frequencies ($<20\%$) in the tumor tissue. Whilst Sanger sequencing suffers from limited sensitivity in the context of detecting rare somatic mutations, quantitative PCR methods have limited scalability for multiplexing which could allow detection of up to hundreds or thousands of mutations in a cost- and time- effective manner. By contrast, by generating digital data, NGS technologies fulfill both these key requirements for their application in the diagnostic setting. Amplicon multiplexing is performed such that thousands of amplicons containing many different mutations are amplified and sequenced together, and sequencing depth is increased to enhance somatic mutation detection; as rare mutations can be detected at the level of single individual sequence reads (Table 1). Cost effectiveness may be further improved by sample indexing whereby multiple samples can be pooled together in the same sequencing run [11,12].

Table 1: Comparison of the common methods to detect rare somatic mutations

Methods	**Sensitivity to detect rare mutations**	**Multiplexing of targets (Amplicons and Mutations)**
Sanger sequencing	Limited	Limited scalability
Array genotyping	Limited (analog data)	High (thousands of mutations)
qPCR	High	Limited scalability
NGS	High (digital data)	High (and cost effective)

The arrival of bench-top sequencers such as Illumina MiSeq and the Life Technologies Ion Torrent Personal Genome Machine (PGM) has further accelerated the pace of NGS transfer into the diagnostic setting. The bench-top sequencers are rather more suitable than the conventional high-throughput sequencing platforms such as Illumina HiSeq or Life Technologies 5500xl SOLiD System in terms of their adoption in the diagnostic setting. The reasons are manifold including the fact that the size of the targeted gene region is very much smaller than that required for either WES or WGS, and the requirement for smaller numbers of clinical samples to be run on a regular basis, which would hamper the batching of larger sample sizes. This leads to a requirement for a lower sequencing throughput in the diagnostic setting. In addition, the sequencing run time for bench-top sequencers is much more rapid, taking several hours rather than the days required for high-throughput NGS platforms [13]. Below we elaborate the challenges and preconditions to bring NGS into a diagnostic setting for therapeutic decision-making and highlight the first CE-IVD NGS diagnostic test for oncology that has been brought to the market.

Although the gradual permeation of NGS into the diagnostic setting has been envisaged, there have been many challenges particularly in the context of detecting somatic mutations in oncology [14-17]. The challenges range from assay design to sample preparation, data analysis, variant calling, and automation of the entire workflow from samples to results. In terms of the assay design, it is critical to ensure that targets are selected so as to be both of clinical relevance and actionable. In addition, the tests should achieve the required levels of analytical and clinical sensitivity and specificity for diagnostic application.

The analytical sensitivity and specificity can be established by detecting known mutations in well-characterized reference materials. On the other hand, clinical performance is established by comparing the concordance and discordance rates of mutation detection in clinical or reference samples already tested by reference methods e.g. quantitative PCR and Sanger sequencing (the latter for mutations with frequencies >20%). As NGS tests are characterized by their high multiplexing capability (i.e. up to hundreds or thousands of mutations can be targeted), testing every single mutation to assess clinical performance represents quite a challenge. What hampers this validation is that a large number of samples would be required to establish the parameters and it is probably impracticable to obtain an adequate number of samples, especially for low population frequency mutations among cancer patients.

In addition to the selection of clinically relevant targets, and the requirements of analytical and clinical performance, an automated workflow from sample preparation to data analysis, variant calling and result reporting is also very critical in the diagnostic setting. The importance of this requirement is well exemplified in the NGS tests, where the workflow involves many steps from DNA extraction from tumor tissue to library construction, template preparation, sequencing and data analysis. Validation is needed to ensure the reproducibility of the workflow, and automation is a critical component for this. For example, robotic automation should ensure reproducibility in terms of quality and quantity of the DNA extraction from tumor tissue and library preparation as well as minimizing operator errors during the tedious manual steps. Automation also brings other significant advantages i.e. reducing the labor requirement for the work (i.e. improving the laboratory's manpower efficiency) and avoiding the need to repeat samples due to operator errors (increasing cost effectiveness). Other critical requirements in the diagnostic workflow lie in ensuring sample traceability, where automatic sample ID list is being transferred from one part of the workflow to another, this serves to avoid manual entry errors in the ID list.

Another challenge that hinders the widespread application of NGS in the diagnostic context is data analysis, which requires specialist expertise in bioinformatics. This includes the optimization of analysis parameters, alignment of sequence reads, variant calling and annotation. This is important to ensure high sensitivity and specificity with respect to detecting the target mutations. Building a robust and validated bioinformatic analysis pipeline to meet the strict requirements of analytical and clinical performance is resource intensive. Furthermore, the analysis part should also be user friendly so that laboratory personnel without specialist bioinformatics training can perform it. The reports generated at the end of the analysis must also be designed specifically to meet requirements in the clinical context, where sufficient QC information must be included to allow assessment of the quality of the runs and samples, whilst the report for clinicians must be concise, summarizing only the mutations being sought or detected.

Although many commercial companies have been working towards bringing NGS oncology tests onto the market, Vela Diagnostics was the first by dint of launching its NGS oncology test and the automated workflow for melanoma with CE-IVD approval in September 2014 (www.veladx.com), and oncology tests for colorectal cancer, non-small cell lung cancer, and thyroid cancer have also been launched subsequently with CE-IVD mark, and all these tests have also obtained TGA (Therapeutic Goods Administration) approval in Australia. These tests are designed to detect somatic mutations in tumor samples aiding the prescription of molecular targeted drugs. This is an important milestone in the translational effort to bring NGS oncology test into the diagnostic setting. The workflow that was launched by Vela Diagnostics to support the NGS oncology tests (http:// www.veladx.com/products.html#ngs) fulfilling the preconditions discussed above bringing NGS tests into the diagnostic setting become a reality. The workflow begins with automated genomic DNA extraction and library construction on a robotic liquid handling system (Figure 1). This automated workflow offers several significant advantages. For example, it ensures the reproducibility of libraries in terms of quality and quantity. Further, in the validated workflow, no QC steps are required to measure the concentration of extracted genomic DNA, or to assess the library's quality and quantity. After automated library construction is completed, the pool library is used for subsequent template preparation and sequencing to generate the sequencing data. The data are analyzed by the software i.e. Sentosa SQ Reporter automatically, where the steps such as alignment and variant calling take place. The automation of the entire workflow also makes it feasible to achieve a turn-around time of 2 days from DNA extraction to report generation for each clinical sample.

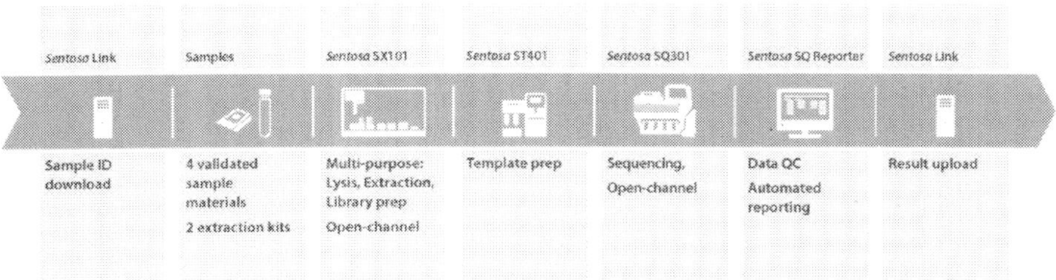

Figure 1: Sentosa NGS Workflow integrated with Sentosa Link. The workflow begins with the generation of a sample ID list using the Sentosa Link. The ID list is then transferred throughout the whole workflow by the Sentosa Link. The DNA extraction from FFPE samples and library construction steps are fully automated on SX101 (with an external PCR amplification step). This is then followed by template preparation and sequencing on the ST401 and SQ301 devices, respectively. Finally, the data are analyzed by SQ Reporter for variant calling, and the reports generated are sent back to Sentosa Link upon electronic sign-off.

The widespread application of NGS tests in the diagnostic setting is growing apace. In addition to oncology tests designed to identify somatic mutations for therapeutic decision-making, NGS also has huge potential as a diagnostic tool in the context of Mendelian disorders. This is well exemplified by the Illumina MiSeqDx Cystic Fibrosis System, which is the first FDA-cleared IVD NGS platform for the comprehensive testing of cystic fibrosis [18]. These tests have so far used a targeted gene sequencing approach to sequence specifically those gene regions containing clinically relevant mutations. It is foreseeable that more NGS diagnostic tests will be developed and launched into the marketplace as the rapidly evolving discipline of genomic medicine is devoted to utilizing genomic information to further the clinical care of patients and to improve health outcomes.

REFERENCES

1. Margulies M, Egholm M, Altman WE, Attiya S, Bader JS, et al. (2005) Genome sequencing in microfabricated high-density picolitre reactors. Nature437: 376-380.
2. Mizzi C, Peters B, Mitropoulou C, Mitropoulos K, Katsila T, et al. (2014)Personalized pharmacogenomics profiling using whole-genome sequencing. Pharmacogenomics 15: 1223-1234.
3. Rabbani B, Tekin M, Mahdieh N (2014) The promise of whole-exome sequencing in medical genetics. J Hum Genet59: 5-15.
4. Ku CS, Cooper DN, Polychronakos C, Naidoo N, Wu M, et al. (2012) Exome sequencing: dual role as a discovery and diagnostic tool. Ann Neurol71: 5-14.
5. Manolio TA, Chisholm RL, Ozenberger B, Roden DM, Williams MS, et al. (2013) Implementing genomic medicine in the clinic: the future is here. Genet Med 15: 258-267.
6. Green ED, Guyer MS (2011) National Human Genome Research Institute: Charting a course for genomic medicine from base pairs to bedside. Nature470: 204-213.
7. Ley TJ, Mardis ER, Ding L, Fulton B, McLellan MD, et al. (2008) DNA sequencing of a cytogenetically normal acute myeloid leukaemia genome. Nature456: 66-72.
8. Worthey EA, Mayer AN, Syverson GD, Helbling D, Bonacci BB, et al. (2011) Making a definitive diagnosis: successful clinical application of whole exome sequencing in a child with intractable inflammatory bowel disease. Genet Med 13: 255-262.
9. Gillis NK, Patel JN, Innocenti F (2014) Clinical implementation of germ line cancer pharmacogenetic variants during the next-generation sequencing era. ClinPharmacolTher95: 269-280.
10. de Mello RA, Madureira P, Carvalho LS, Araújo A, O'Brien M, et al. (2013) EGFR and KRAS mutations, and ALK fusions: current developments and personalized therapies for patients with advanced non-small-cell lung cancer. Pharmacogenomics 14: 1765-1777.
11. Mwenifumbo JC, Marra MA (2013) Cancer genome-sequencing study design. Nat Rev Genet 14: 321-332.
12. Meyerson M, Gabriel S, Getz G (2010) Advances in understanding

cancer genomes through second-generation sequencing. Nat Rev Genet 11: 685-696.

13. Ku CS, Wu M, Cooper DN, Naidoo N, Pawitan Y, et al. (2012) Technological advances in DNA sequence enrichment and sequencing for germline genetic diagnosis. Expert Rev Mol Diagn 12: 159-173.
14. Bennett NC, Farah CS (2014) Next-generation sequencing in clinical oncology: next steps towards clinical validation. Cancers (Basel) 6: 2296-2312.
15. Salto-Tellez M, Gonzalez de CD (2014) Next-generation sequencing: a change of paradigm in molecular diagnostic validation. J Pathol 234: 5-10.
16. Pant S, Weiner R, Marton MJ (2014) Navigating the rapids: the development of regulated next-generation sequencing-based clinical trial assays and companion diagnostics. Front Oncol4: 78.
17. Xuan J, Yu Y, Qing T, Guo L, Shi L (2013) Next-generation sequencing in the clinic: promises and challenges. Cancer Lett340: 284-295.
18. Bijwaard K, Dickey JS, Kelm K, TeÅ¾ak Å½ (2015) The first FDA marketing authorizations of next-generation sequencing technology and tests: challenges, solutions and impact for future assays. Expert Rev Mol Diagn 15: 33-40.

CHAPTER 15

NEXT GENERATION SEQUENCING IN CONGENITAL HEART DISEASE: GENE DISCOVERY AND CLINICAL APPLICATION

Gillian M. Blue[1,2,3] and David S. Winlaw[1,2,3]

[1]Kids Heart Research, The Children's Hospital at Westmead, Sydney, Australia

[2]Heart Centre for Children, The Children's Hospital at Westmead, Sydney, Australia

[3]Sydney Medical School, University of Sydney, Australia

ABSTRACT

Congenital Heart Disease (CHD) affects 6-8 per 1000 live births. It is the most common cause of death in newborn infants and is a significant burden on society with an increasing adult population. A number of genes have been implicated in disease development; however traditional research techniques such as linkage analysis are generally not suited as they require large families with multiple affected individuals which are rare in CHD. While more recent techniques, such as GWAS, provide insight into population-attributable risks, the information cannot be directly related to the affected individual. In comparison, the information gained from next generation sequencing (NGS) technology can directly impact the treatment and care of patients and their families. For this reason NGS is rapidly becoming the analysis of choice in both clinical and research arenas. The different types of NGS techniques have unique sets of advantages and disadvantages and it is important to assess these in the context of the disease characteristics, as well as the purpose of the analysis, so that the most appropriate technology is selected. Based on our experience and recent findings from research in CHD, we propose that a gene panel-based approach is currently more suited to clinical workflow and the molecular analysis of CHD and that exome sequencing and whole genome sequencing are best suited to a research environment that can provide the necessary support and analyses required for novel gene discovery.

Keywords

Next generation sequencing, Gene panels, Congenital heart disease, Molecular diagnosis

INTRODUCTION

Congenital Heart Disease (CHD) affects 6-8 per 1000 live births [1]. It is the most common cause of death in newborn infants in the western world and is also a significant cause of fetal loss. A third of live-born infants will require surgery within the first year of life; some requiring many more surgeries throughout adulthood. Due to improvements in surgical techniques and treatment, it is now generally accepted that adult CHD constitutes the majority of the CHD population in high-income countries [2]. CHD places a significant burden on the affected individual and their families with numerous studies reporting impairments in psychosocial functioning consistent with clinical criteria for depression, anxiety and somatisation [3-5]. CHD also places a significant burden on the community both in terms of providing care and funding for treatment.

EARLY RESEARCH TECHNIQUES

The high heritability of CHD (0.6-0.7) suggest a strong genetic component and current research points to both genetic and environmental contributions to disease development in the majority of cases, which tend to be sporadic in nature. CHD is a complex disease in that it often displays genetic heterogeneity, variable penetrance and variable expression. Approximately 20% of cases present as part of a genetic syndrome or chromosomal abnormality and for these individuals the 'reason' or 'cause' for their CHD is generally understood [6]. A small portion of cases exhibit familial forms of disease with seemingly Mendelian inheritance patterns. Most of the genetic information we know about CHD to date has been obtained from studying familial forms of disease due to their suitability to early research techniques, including linkage analysis and candidate gene studies [7-10]. While linkage analysis is a powerful tool and has successfully uncovered novel disease genes in many genetic conditions, it does have limitations. A prerequisite to using this technique is large families with multiple affected individuals. Linkage also requires the disease to segregate according to Mendelian principles with complete penetrance. Linkage analysis is therefore not ideal for studies on complex diseases such as CHD, as large families with multiple affected individuals, segregating disease according to Mendelian principles are rare.

RECENT RESEARCH TECHNIQUES

More recent techniques, such as genome-wide association studies (GWAS), have provided evidence that common genetic variation can influence the risk of certain types of CHD [11-13]. However, while GWAS provide insight into population-attributable risks, the information is not directly transferable to the individual with a heart defect. Advances in sequencing technology, combined with the steady decline in costs, have resulted in the development of new-age sequencing technologies that are collectively termed next generation sequencing (NGS). The techniques can be targeted (exome sequencing and disease-specific gene panels), or non-targeted (whole genome sequencing (WGS). While all types of NGS use massively parallel sequencing, the targeted approaches include a prior capture and enrichment step ensuring only targeted regions are sequenced.

NGS is well suited to the study of complex, heterogeneous diseases such as CHD. First and foremost, and in contrast to GWAS, the results of NGS are directly applicable to the affected individual. This information has the potential to provide diagnostic prospects and directly impact the treatment and care of patients and their families. Secondly, the technology enables the rapid simultaneous interrogation of multiple genes, or even the entire genome in the case of WGS, which is particularly appealing to the study of heterogeneous diseases. Finally, NGS is (in most cases) unbiased in that it permits the discovery of both known and novel variants and disease genes, thereby promoting gene discovery.

EXOME SEQUENCING IN CHD

A number of studies have successfully applied NGS to CHD, more specifically exome sequencing. Zaidi et al. identified a significant increase in *de novo* coding variants in genes expressed in the developing heart in over 300 sporadic cases of CHD [14]. The study was important in defining the mechanisms of causation with a higher burden of variation evident in genes involved in cardiac development. However, while these findings contribute significantly towards the overall understanding of CHD and developmental pathways involved, they did not provide a cause of CHD for the majority of families included in this study.

Other studies have focused on individual families with presumed Mendelian forms of CHD [15-20] **Table 1**. Exome sequencing in families exhibiting Mendelian inheritance assists greatly with analysis as the presumed mode of inheritance can be used as part of the filtering strategy. In this way, exome sequencing successfully identified pathogenic variants in most of these families; however, the evidence on whether the variant identified was solely responsible for the CHD, is questionable in some cases. Together, these studies highlight both the successes as well as the complexities involved in applying exome sequencing to the study of complex, heterogeneous CHD. Despite sophisticated study designs and filtering strategies, reduced penetrance and variable expression present significant hurdles for analysis. The biggest obstacle during any exome analysis is to single out the causal variant from the thousands of variants identified during sequencing. In order to assist with this, strategies are applied to filter out the unlikely variants; however, these strategies are based upon a number of assumptions, including the assumption that the causal variant is fully penetrant. This may not always be the case in CHD, despite the family appearing to be 'Mendelian'. Familial forms of CHD are known to display variable expression, various ages of presentation as well as variable penetrance [21] which complicate the exome sequencing analysis. Furthermore, in any exome analysis, the filtering process almost always contains a degree of bias in that it is limited to what is known; whether this is filtering against variants in known databases, retaining variants deemed to be pathogenic following *in silico* analyses or during educated variant curation.

Table 1: Summary of the available literature on exome sequencing in familial forms of CHD

Study	CHD type	Sample size	Suspected inheritance	Analysis type	Likely causal variant identified	Novel/ Known variant	Segregation with disease	In silico predictions
Arrington et al.	ASD, VSD, Coarc	2	Dominant	ES	MYH6 (p.Ala290Pro)	Known	No	Damaging
Liu et al.	ASD	1	Dominant	ES and CHD gene filter	TBX20 (p.Asp176Asn)	Novel	Yes	Damaging
Ta-Shma et al.	TA	1	Recessive	ES and homozygosity mapping	PLXND1 (p.Arg1299Cys)	Novel	Yes*	Damaging
Greenwa et al.	ASD	2	Dominant	ES	ACTC1 (p.Met178Leu)	Novel	Yes	Damaging and Benign
Martin et al.	BAV, VSD, AVSD	17	Dominant	ES	No	n/a	No	n/a
Tariq et al.	L-TGA, heterotaxy	1	Dominant	ES andhomozygosity mapping	SHROOM3 (p.Gly60Val)	Novel	Unknown#	Damaging

*No other affected individuals available for testing; # Only proband tested. ASD: Atrial Septal Defect; AVSD: Atrioventricular Septal Defect;

BAV: Bicuspid Aortic Valve; CHD: Congenital Heart Disease; Coarc: Coarctation of the Aorta; ES: Exome Sequencing; L-TGA: Congenitally Corrected Transposition of the Great Arteries; n/a: not applicable; TA: Truncus Arteriosus; VSD: Ventricular Septal Defect

NGS technology, more specifically exome sequencing and WGS, have some disadvantages that complicate their use in general, particularly in the clinical setting. From a technical perspective, issues with coverage, both in terms of depth as well as capture of on-target regions, mean that not all regions of the exome or genome have been covered to an appropriate depth, if at all. It is therefore possible that a causal variant is located in an uncovered, or poorly covered, region and will be missed [22]. Exome sequencing on-target coverage is generally reported in the range of 95%-97% [23] which is a concern for clinicians as causal variants may be missed if they fall in regions of poor, or no, coverage. A recent study evaluated the exome sequencing coverage of clinically relevant cardiac genes, including 31 genes present on the American College of Medical Genetics (ACMG) list of genes for reporting of incidental findings [24] as well as an additional 19 genes related to CHD [25]. They found that there was considerable variation in coverage of the individual genes; with 10 genes, including six from the ACMG list, covered less than the clinically required 30x.

Another concern is the lengthy data analyses and the issues with data storage, particularly in regards to cost and the resources required. These issues hinder efficient translation into a clinical setting.

Finally, there is also the issue of 'incidental findings'. Sequencing an individual's entire coding genome means that in some cases variants associated with a disease other than the one in question may be identified. For this reason, lengthy discussions about this possibility need to be conducted with participants or patients to ensure their preferences are met should an 'incidental finding' be identified. This can be very time-consuming and may result in significant delays, particularly from a research perspective.

Gene Panels in CHD

There is no question about the power that exome sequencing and WGS offer the investigator in terms of novel gene discovery due to their unbiased approach; however issues relating to coverage, analysis and storage of large amounts of data, as well as reporting of incidental findings complicate their use in the clinical setting [26]. In comparison, limiting the capture regions to only those known to be associated with the disease(s) of interest mitigates some of these issues; however this is at the cost of identifying novel disease genes.

Specific tools enable the design of individualized gene panels so that parameters can be individually adjusted to ensure maximum coverage of on-target regions. The substantial reduction in sequencing of a gene panel also allows for a significantly increased coverage depth compared with exome sequencing. This is usually ten-fold, or more, higher than the average coverage achieved using exome sequencing [26,27]. A higher depth of coverage significantly reduces the false positive rate, resulting in higher confidence in variant calling. In some areas small regions of the target genes may; however, still be missed. This is usually due to probe failure or issues with regions of high GC content. Being able to supplement these small regions with Sanger sequencing further ensures that all target regions have been adequately covered and increases confidence with results.

Gene panels also produce substantially less data for analysis and storage, making them an arguably more suitable approach in the diagnostic arena [28]. Furthermore, issues relating to reporting of incidental findings can be avoided entirely as only disease-relevant genes are screened.

CONCLUSION

With the steady decline in sequencing costs, exome sequencing has entered the clinical arena and WGS will, no doubt, follow suit in the coming years. We have experience with both exome sequencing (unpublished work) and gene panels in CHD [29] and our findings suggest that at present, a gene panel-based approach is more suited to the clinical and molecular analysis of CHD, at least as an initial screen, followed by exome sequencing in the event that no candidate variants are identified. Exome sequencing and WGS, on the other hand, are currently better suited to a research environment where novel gene discovery can be supported with appropriate functional studies to confirm and substantiate findings. Our experience is; however, limited to CHD, which is a complex disease, so this may not be the case for other genetic diseases with 'simpler' inheritances. Exome sequencing analysis is more straightforward in true Mendelian forms of disease where complicating factors such as incomplete penetrance and possible di- , oligo- or polygenic inheritances can be avoided. Furthermore, being able to identify both known and novel variants, using exome sequencing and WGS cannot be underestimated, especially in gene discovery research.

In summary, NGS technologies are revolutionizing the field of genetics and genetic medicine. While the different techniques (WGS, exome sequencing and gene panels) use the same sequencing processes, they differ in their capture methods and therefore each present a unique set of advantages and disadvantages. For this reason, the type of NGS technology applied to a genetic condition, needs to be carefully evaluated both in the context of the disease characteristics, as well as the purpose of the analysis i.e. whether it is clinical/diagnostic or research driven.

REFERENCES

1. Hoffman JI, Kaplan S (2002) The incidence of congenital heart disease. J Am Coll Cardiol 39:1890-1900.
2. Marelli AJ, Ionescu-Ittu R, Mackie AS, Guo L, Dendukuri N, et al. (2014) Lifetime Prevalence of Congenital Heart Disease in the General Population From 2000 to 2010. Circulation130:749-756.
3. Lawoko S, Soares JJ (2002) Distress and hopelessness among parents of children with congenital heart disease, parents of children with other diseases, and parents of healthy children. J Psychosom Res52:193-208.
4. Kasparian NA, Fidock B, Sholler GF, Camphausen C, Murphy DN, et al. (2014) Parents' perceptions of genetics services for congenital heart disease: the role of demographic, clinical, and psychological factors in determining service attendance. Genet Med16:460-468.
5. Blue GM, Kasparian NA, Sholler GF, Kirk EP, Winlaw DS (2015) Genetic counselling in parents of children with congenital heart disease significantly improves knowledge about causation and enhances psychosocial functioning. Int J Cardiol178:124-130.
6. Blue GM, Kirk EP, Sholler GF, Harvey RP, Winlaw DS (2012) Congenital heart disease: current knowledge about causes and

inheritance. Med J Aust 197:155-159.

7. Garg V, Kathiriya IS, Barnes R, Schluterman MK, King IN, et al. (2003) GATA4 mutations cause human congenital heart defects and reveal an interaction with TBX5. Nature424:443-447.
8. Schott JJ, Benson DW, Basson CT, Pease W, Silberbach GM, et al. (1998) Congenital Heart Disease Caused by Mutations in the Transcription Factor NKX2-5. Science281:108-111.
9. Bonnet D, Pelet A, Legeai-Mallet L, Sidi D, Mathieu M, et al. (1994) A gene for Holt-Oram syndrome maps to the distal long arm of chromosome 12. Nat Genet 6:405-408.
10. Basson CT, Cowley GS, Solomon SD, Weissman B, Poznanski AK, et al. (1994) The Clinical and Genetic Spectrum of the Holt-Oram Syndrome (Heart-Hand Syndrome). N Engl J Med 330:885-891.
11. Stevens KN, Hakonarson H, Kim CE, Doevendans PA, Koeleman BPC, et al.(2010) Common Variation in ISL1 Confers Genetic Susceptibility for Human Congenital Heart Disease. PLoS ONE 5:e10855.
12. Cordell HJ, Bentham J, Topf A, Zelenika D, Heath S, et al.(2013) Genome-wide association study of multiple congenital heart disease phenotypes identifies a susceptibility locus for atrial septal defect at chromosome 4p16. Nat Genet45:822-824.
13. Cordell HJ, Töpf A, Mamasoula C, Postma AV, Bentham J, et al. (2013) Genome-wide association study identifies loci on 12q24 and 13q32 associated with Tetralogy of Fallot. Hum Mol Genet 22:1473-1481.
14. Zaidi S, Choi M, Wakimoto H, Ma L, Jiang J, et al.(2013) De novo mutations in histone-modifying genes in congenital heart disease. Nature 498:220-223.
15. Arrington CB, Bleyl SB, Matsunami N, Bonnell GD, Otterud BEM, et al. (2012) Exome Analysis of a Family With Pleiotropic Congenital Heart Disease / Clinical Perspective. Circulation: Cardiovascular Genetics 5:175-182.
16. Liu C, Shen A, Li X, Jiao W, Zhang X, et al. (2008) T-box transcription factor TBX20 mutations in Chinese patients with congenital heart disease. Eur J Med Genet 51:580-587.

17. Ta-Shma A, Pierri CL, Stepensky P, Shaag A, Zenvirt S, et al. (2013) Isolated truncus arteriosus associated with a mutation in the plexin-D1 gene. American Journal of Medical Genetics Part A. 161:3115-3120.
18. Greenway SC, McLeod R, Hume S, Roslin NM, Alvarez N, et al.(2014) Exome Sequencing Identifies a Novel Variant in ACTC1 Associated With Familial Atrial Septal Defect. Can J Cardiol. 30:181-187.
19. Martin LJP, Pilipenko VP, Kaufman KMP, Cripe LMD, Kottyan LCP, et al. (2014) Whole Exome Sequencing for Familial Bicuspid Aortic Valve Identifies Putative Variants. Circulation: Cardiovascular Genetics 7:677-683.
20. Tariq M, Belmont J, Lalani S, Smolarek T, Ware S (2011) SHROOM3 is a novel candidate for heterotaxy identified by whole exome sequencing. Genome Biol 12:R91.
21. Kirk EP, Sunde M, Costa MW, Rankin SA, Wolstein O, et al. (2007) Mutations in Cardiac T-Box Factor Gene TBX20 Are Associated with Diverse Cardiac Pathologies, Including Defects of Septation and Valvulogenesis and Cardiomyopathy. The American Journal of Human Genetics. 81:280-291.
22. Gilissen C, Hoischen A, Brunner HG, Veltman JA (2012) Disease gene identification strategies for exome sequencing. Eur J Hum Genet 20:490-497.
23. Grody WW, Thompson BH, Hudgins L (2013) Whole-Exome/ Genome Sequencing and Genomics. Pediatrics 132:S211-S5.
24. Green RC, Berg JS, Grody WW, Kalia SS, Korf BR, et al. (2013) ACMG recommendations for reporting of incidental findings in clinical exome and genome sequencing. Genet Med. 15:565-574.
25. Manase D, D'Alessandro LC, Manickaraj AK, Al Turki S, Hurles ME, et al. (2017) High throughput exome coverage of clinically relevant cardiac genes. BMC Med Genomics 7:10.
26. Sikkema-Raddatz B, Johansson LF, de Boer EN, Almomani R, Boven LG, et al.(2013) Targeted Next-Generation Sequencing can Replace Sanger Sequencing in Clinical Diagnostics. Hum Mutat 34:1035-1042.
27. Meder B, Haas J, Keller A, Heid C, Just S, et al. (2011) Targeted

Next-Generation Sequencing for the Molecular Genetic Diagnostics of Cardiomyopathies / Clinical Perspective. Circulation: Cardiovascular Genetics 4:110-122.

28. Rehm HL (2013) Disease-targeted sequencing: a cornerstone in the clinic. Nat Rev Genet 14:295-300.

29. Blue GM, Kirk EP, Giannoulatou E, Dunwoodie SL, Ho JWK, et al.(2014) Targeted Next-Generation Sequencing Identifies Pathogenic Variants in Familial Congenital Heart Disease. J Am CollCardiol64:2498-2506.

CHAPTER 16

GENOME WIDE ASSOCIATION STUDIES AND NEXT GENERATION SEQUENCING TECHNOLOGIES IN OSTEOARTHRITIS

Thomas Mabey and Sittisak Honsawek

Department of Biochemistry, Faculty of Medicine, Chulalongkorn University, King Chulalongkorn Memorial Hospital, Thailand

ABSTRACT

Osteoarthritis is a commonly age-related degenerative joint disorder characterized by articular cartilage degradation, subchondral bone sclerosis, osteophyte formation, and synovial membrane inflammation. Various environmental, biomechanical, and genetic factors have been recognized as playing essential roles in OA development. A number of studies have endeavored to decipher the pathogenesis of osteoarthritis. In an attempt to identify the genetic markers of complex diseases such as osteoarthritis, there has been a paradigm shift away from traditional linkage mapping studies and candidate gene association studies to higher-density genome-wide association studies. This introduction to genome-wide association studies and next-generation sequencing technologies provides of these areas and then considers their relevance to in osteoarthritis. High-throughput genomic and transcriptomic methods have resulted in a paradigm shift in the way osteoarthritis is perceived and have changed the way translational research is performed. This review presents an overview of high-throughput genome wide association and next generation sequencing in osteoarthritis and discusses clinical applications of these technologies.

Keywords

Genome wide association studies, next generation sequencing, osteoarthritis

INTRODUCTION

Osteoarthritis (OA) is a progressive degenerative joint disease that is characterized by articular cartilage loss, osteophyte formation, subchondral bone sclerosis, and synovial inflammation. OA is associated with joint space narrowing and osteophyte formation leading to loss of joint function and pain. Whilst the exact aetiology of the disease remains unclear, age, gender, obesity, previous joint trauma, menopausal status, and genetic variations are known risk factors associated with OA [1].

A wide range of biological processes play important roles in the pathogenesis of OA. Pro- and anti- inflammatory, angiogenic, and growth signaling pathways are central aspects of disease progression. Likewise in other multifactorial disorders, each aspect of the disease is subject to genetic variations in the genes encoding and regulating individual proteins in each pathway resulting in alterations in patient's susceptibility to OA, prognosis, and the efficacy of intervention. Due to the large range of genetic diversity involved, high-throughput sequencing techniques are valuable tools for the continued research in such diseases.

Next generation sequencing (NGS) is a technique that has already begun to revolutionize research and clinical genomics. With the ability to sequence target transcriptomes, exomes or whole genomes in a relatively quick and economically viable way, the understanding of genetic factors in diseases is improving, leading to improved therapeutics and tailored treatment plans in clinical practice. **Figure 1** displays a summary of techniques being used to analyse genetic variations at various levels (reviewed in more details by Soon et al. [2] and Zhou et al. [3]).

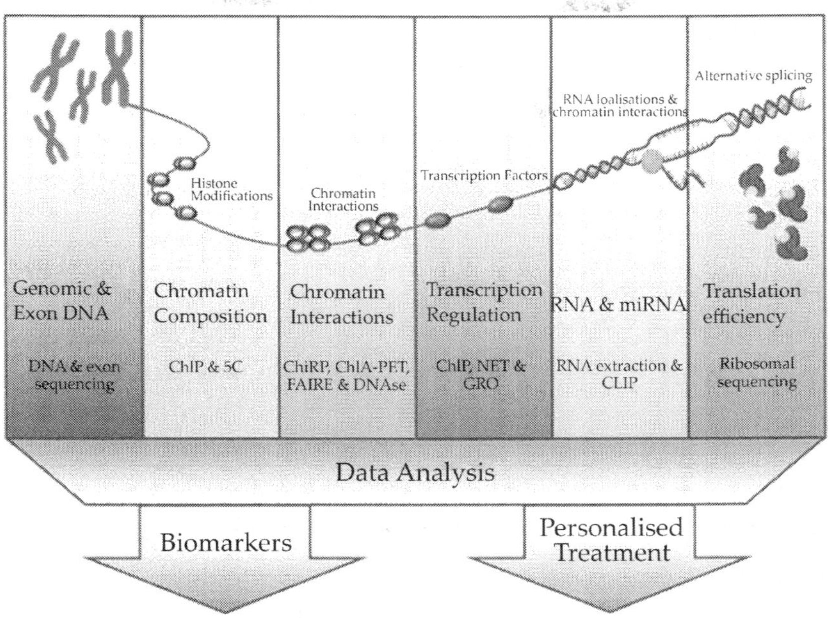

Figure 1: The levels at which genetic variations can occur and example techniques for their analysis. ChIP: Chromatin Immunoprecipitation; 5C: ; ChiRP: Chromatin isolation by RNA purification; ChIA-PET: Chromatin Interaction Analysis with Paired-End Tag Sequencing; FAIRE: ormaldehyde-ssisted solation of egulatory lements; DNAse: DNase I hypersensitive site sequencing; NET: Native elongating transcript sequencing; GRO: Global run-on sequencing; CLIP: Cross-linking immunoprecipitation.

Additionally, developments in computational analysis and the growth of bioinformatic databases, such as HapMap and HuGE, have supported the increased use of large scale genetic studies. Genomic variations in individuals include single nucleotide polymorphisms (SNPs), insertions, and deletions which can lead to the heritability of diseases. NGS offers a genome-wide view of a patient which can prove to be an indispensable tool in multifactorial disease like OA. Here, we briefly describe how NGS is being utilized in OA and how future applications will likely improve our understanding of the pathophysiology and ultimatelypatient care.

CANDIDATE GENE ASSOCIATION STUDIES

A large number of studies in SNPs and their association with OA have provided greater understanding of its pathophysiology in recent years. The range of genes investigated reflects the complex nature of OA. **Table 1** demonstrates the range of the 88 genes associated with knee OA to date. Due to differences in ethnic, age, sex, and other variables, numerous studies remain to be replicated and often have limited statistical power from small sample sizes.

Table 1: The 88 genes associated with knee osteoarthritis according to HuGE Navigator [4]

ACE	DIO2	IL1RN	PRKAR2B*
ADAM12	DIO3	IL6	PTGS2*
AGER	DUS4L*	KL	RHOB
ANP32A	EDG2	LEP	SERPINA3
AR	EPAS1	LEPR	SLC26A2
ASPN	ESR1	LOC344875	SMAD3
BCAP29*	ESR2	LPAR1	SOST
BDKRB2	FRZB	LRCH1*	TLR2
BMP2	GDF5	LRP5	TLR3
BTNL2*	GHRHR	LRP6	TLR4
CALCA	GPR22**	MCF2L	TLR7
CALM1	HFE	MMP1	TLR8
CCL2	HLA-DQB1*	MMP13	TLR9
CD36	HLA-DRB1	MMP2	TNF
CILP	HPBP*	MMP3	TNFAIP6
CLEC3B	ICAM1	MMP9	TNFRSF11B
COG5**	IL10	NCOR2	TNFRSF1A
COL11A1	IL18	PAPSS2	TNFRSF1B
COL2A1	IL18R1	PCSK6	TRPV1
COL6A4P1	IL1A	PITX1	TXNDC3
COL9A3	IL1B	PLA2G4A*	VDR
COMP	IL1R1	PPARG	VEGFA

* Gene association supported by a GWAS [5-9] ** Gene association supported by more than one GWAS [10, 11]

However, they have offered an inside into a range of genetic variations and their effects in OA. Frequently genes have multiple possible SNP locations in the exons, introns, or promotor and regulatory regions which can complicate the roles possible mutations may play. The narrow field of view of such studies limits their ability to efficiently identify disease-associated variations but do allow future studies build on their findings.

GENOME WIDE ASSOCIATION STUDIES

By scanning the entire genome of large samples of individuals with or without a disease, variations can be found that may be associated with a disease or condition. Such high-throughput studies rely on large sample sizes to increase the statistical power of their results and are made possible by NGS. Hitherto, 9 genome wide association studies (GWASs) have been conducted with respect to OA [12]. Panoutsopoulou and Zeggini have reviewd GWASs of European and Asian individuals [13]. A total of 15 loci have been identified through the use of GWASs. The authors also noted the importance of phenotype definition. Qualitative techniques are the primary diagnostic tools for OA. The Kellgren-Lawrence scale [14] is used extensively as a measure of disease severity in osteoarthritis but it is subjective and open to interpretation in clinical practice. Such variables in the assessment of osteoarthritis severity introduce uncertainties in the results of GWASs which should be considered when interpreting their conclusions.

The arcOGEN GWAS study revealed 5 novel OA-associated loci and a further three 3 just below the genome-wide association threshold in a sample group of 7410 individuals [15]. The rs6976 SNP in the *GNL3* gene on chromosome 3 showed the strongest association. Guanine nucleotide binding protein-like 3, also known as nucleostemin, is encoded by the *GNL3* gene and is known to regulate the cell cycle in mesenchymal stem cells which ultimately differentiate into cartilage forming chondrocytes [16]. However, its role in OA is not fully understood. Whilst this study was critisised by Pang et al. for having beem restricted to European participants [17], these findings provide the basis of future functional studies.

The use of meta-analysis of multiple cohorts improves the statistical power of such studies. Very recently, Rodriguez-Fontenla et al. performed a meta-analysis of 9 SNP-level GWASs [18]. They found that of 199 previous suggested candidate genes associated with OA, only 2 showed significant association, *COL11A1* and *VEGF*. Vascular endothelial growth factor (VEGF) levels in plasma synovial fluid have previously been reported to correlate with disease severity in knee osteoarthritis [19].

With thousands of participants, GWASs yield vast amounts of information which presents possible problems with information security and patient privacy. As DNA is the most personal of information types, study participants should be well informed of the risks and all precautions taken to prevent data loss or misuse.

Although the large sample sizes and often complex data analysis can be expensive and time consuming, GWASs are powerful tools at detecting variations on a genome-wide scale which can serve as the bedrock for further researches in cell and molecular biology.

BIOMARKERS AND THERAPEUTICS

The search for diagnostic and disease-assessment markers in OA has been primarily focused on investigating biochemical components of cartilage degradation and inflammation related cytokines in serum/plasma and synovial fluid [20]. The potential of biomarkers to assist in diagnosis of OA is obviously an attractive tool for clinicians. However, local and systemic biochemical levels have limited use for predicting the susceptibility of osteoarthritis and other diseases with long onset periods often lasting many years. Genetic variations that can predict the risk of such disease may help patients and clinicians take mitigating or preventative measures. Additionally, the development of effective therapeutics may be supported by improved understanding of the pathophysiology of OA that comes from a greater knowledge of the genetic influences.

Genetic markers from a range of physiological processes have been investigated. The IL-18 rs1946518 SNP has been shown to distinguish between knee OA patients and healthy controls with borderline significance [21]. Interleukin (IL)-18 is a proinflammatory cytokine which has been shown association with disease progression [22]. Moreover, SNPs of the osteopontin gene which is involved in bone formation and remodeling have been associated with an increased susceptibility of OA and radiographic KL grades [23]. Our laboratory recently built on the work of Rodriguez-Lopez et al. and showed that a non-synonymous (ns) SNP at rs4747096 in ADAMTS14 to be associated with knee OA in Thai and Caucasian females [24,25]. However, this study also shows that genetic variations can be heavily gender-dependent and therefore, the sex of individuals should be considered when performing genetic studies.

By understanding the variations in genes and the resultant effects, therapeutic agents can be designed to target these vulnerabilities. In the future, as the understanding of genetic OA continues to develop, gene therapy may be a viable treatment method in OA. Further research in this field is ongoing but studies show promising results, however, therapy may be most effective in early and often asymptomatic stages of the disease which may be a hurdle in the clinical application of such treatments [26].

Sanger sequencing (termination sequencing) and PCR techniques lack a wide field of view when investigating a patient's genetics. Furthermore, shortcomings in the older techniques often result in rarer deletion mutations being missed. As shown by Koboldt et al. [27], such mutations can affect protein structure and function leading to changes in the efficacy of small molecule drugs. Despite the introduction of NGS, PCR and Sanger sequencing still have roles to play in the development of better understanding of the genetic roles in OA. NGS offers a wider perspective which results in less bias and open the possibility to investigate many genes simultaneously and at a lower cost.

Due to OA being a multifactorial disease, the assessment of OA patients will require the ability to access multiple genes for multiple variations in a short time frame. NGS has the power to revolutionise medicine with personalised treatment courses.

CLINICAL APPLICATIONS

Medicine will be revolutionized by advances towards so called "bedside" sequencing. Future patients may have tailored treatment programs selected by their genetic information suggesting the efficacy of different therapeutics; however the risk of false data is always a concern. In OA, antibodies have been used in clinical trials to inhibit proinflammatory agents, such as tumour necrosis factor (TNF)-α, IL-1α and IL-1β, and matrix metalloproteinase (MMP)-13 [28-30]. Other existing treatments include the administration of hyaluronic acid and non-steroidal anti-inflammatory drugs (NSAIDs). However, the efficacy of these treatments may be affected by phenotypic variations on a cellular and tissue level because of genomic variations in an individual. Being able to predict which treatment is best suited to a patient will not only be beneficial to the patient's care but also save money. As the cost of sequencing continues to fall, the money saved from selective administration will inevitably outweigh the cost of analysis making it an economically viable option in the clinical environment.

CONCLUSION

In conclusion, next generation sequencing is a powerful tool that promises to continue to revolutionize research and clinical genetics. Large scale genome wide association studies offer broad views of individuals with complex disease and enable the detection of associated variations that may otherwise have been missed by older techniques. Additionally, through the relatively quick and cost effective sequencing of patient genomes, tailored medicine has the potential to offer personalized treatment plans and improve patient care.

ACKNOWLEDGEMENTS

This study has been facilitated by the Ratchadapiseksompotch Fund (RA55/22), Faculty of Medicine, Chulalongkorn University.

REFERENCES

1. Suri P, Morgenroth DC, Hunter DJ (2012) Epidemiology of osteoarthritis and associated comorbidities. Pm r 4: S10-19.
2. Soon WW, Hariharan M, Snyder MP (2013) High-throughput sequencing for biology and medicine. MolSystBiol9: 640.
3. Zhou X, Ren L, Meng Q, Li Y, Yu Y, et al. (2010) The next-generation sequencing technology and application. Protein Cell 1: 520-536.
4. Yu W, Gwinn M, Clyne M, Yesupriya A, Khoury MJ (2008) A navigator for human genome epidemiology. In: Nat Genet United States 124-125.
5. Valdes AM, Loughlin J, Timms KM, van Meurs JJ, Southam L, et

al. (2008) Genome-wide association scan identifies a prostaglandin-endoperoxide synthase 2 variant involved in risk of knee osteoarthritis. Am J Hum Genet 82: 1231-1240.

6. Spector TD, Reneland RH, Mah S, Valdes AM, Hart DJ, et al. (2006) Association between a variation in LRCH1 and knee osteoarthritis: a genome-wide single-nucleotide polymorphism association study using DNA pooling. Arthritis Rheum 54:524-532.
7. Nakajima M, Takahashi A, Kou I, Rodriguez-Fontenla C, Gomez-Reino JJ, et al. (2010) New sequence variants in HLA class II/III region associated with susceptibility to knee osteoarthritis identified by genome-wide association study. PLoS One 5: e9723.
8. Valdes AM, Styrkarsdottir U, Doherty M, Morris DL, Mangino M, et al. (2011) Large scale replication study of the association between HLA class II/BTNL2 variants and osteoarthritis of the knee in European-descent populations. PLoS One 6: e23371.
9. Panoutsopoulou K, Southam L, Elliott KS, Wrayner N, Zhai G, et al. (2011) Insights into the genetic architecture of osteoarthritis from stage 1 of the arcOGEN study. Ann Rheum Dis 70: 864-867.
10. Evangelou E, Valdes AM, Kerkhof HJ, Styrkarsdottir U, Zhu Y, et al. (2011) Meta-analysis of genome-wide association studies confirms a susceptibility locus for knee osteoarthritis on chromosome 7q22. Ann Rheum Dis 70: 349-355.
11. Kerkhof HJ, Lories RJ, Meulenbelt I, Jonsdottir I, Valdes AM, et al. (2010) A genome-wide association study identifies an osteoarthritis susceptibility locus on chromosome 7q22. Arthritis Rheum 62: 499-510.
12. Hindorff LA, Sethupathy P, Junkins HA, Ramos EM, Mehta JP, et al. (2009) Potential etiologic and functional implications of genome-wide association loci for human diseases and traits. ProcNatlAcadSci USA 106: 9362-9367.
13. Panoutsopoulou K, Zeggini E (2013) Advances in osteoarthritis genetics. J Med Genet 50: 715-724.
14. Kellgren JH, Lawrence JS (1957) Radiological assessment ofosteo-arthrosis. Ann Rheum Dis 16: 494-502.
15. Zeggini E, Panoutsopoulou K, Southam L, Rayner NW, Day-Williams AG, et al. (2012) Identification of new susceptibility loci

for osteoarthritis (arcOGEN): a genome-wide association study. Lancet 380: 815-823.

16. Ma H, Pederson T (2008)Nucleostemin: a multiplex regulator of cell-cycle progression. Trends Cell Biol18: 575-579.
17. Pang H, Luo F, Dai F, Wu XH, Xu JZ (2013) Genome-wide association study for osteoarthritis. Lancet 381: 372-373
18. Rodriguez-Fontenla C, Calaza M, Evangelou E, Valdes AM, Arden N, et al. (2013) Assessment of osteoarthritis candidate genes in a meta-analysis of 9 genome-wide association studies. Arthritis Rheum.
19. Saetan N, Honsawek S, Tanavalee A, Yuktanandana P, Meknavin S, et al. (2013) Relationship of plasma and synovial fluid vascular endothelial growth factor with radiographic severity in primary knee osteoarthritis. IntOrthop.
20. Lafeber FP, van Spil WE (2013) Osteoarthritis year 2013 in review: biomarkers; reflecting before moving forward, one step at a time. Osteoarthritis Cartilage 21:1452-1464.
21. Hulin-Curtis SL, Bidwell JL, Perry MJ (2012) Evaluation of IL18 and IL18R1 polymorphisms: genetic susceptibility to knee osteoarthritis. Int J Immunogenet 39: 106-109.
22. Denoble AE, Huffman KM, Stabler TV, Kelly SJ, Hershfield MS, et al. (2011) Uric acid is a danger signal of increasing risk for osteoarthritis through inflammasome activation. ProcNatlAcadSci USA 108: 2088-2093.
23. Jiang Y, Yao M, Liu Q, Zhou C (2013) OPN gene polymorphisms influence the risk of knee OA and OPN levels in synovial fluid in a Chinese population. Arthritis Res Ther15: R3.
24. Poonpet T, Honsawek S, Tammachote N, Kanitnate S, Tammachote R (2013) ADAMTS14 gene polymorphism associated with knee osteoarthritis in Thai women. Genet Mol Res 12: 5301-5309.
25. Rodriguez-Lopez J, Pombo-Suarez M, Loughlin J, Tsezou A, Blanco FJ,et al. (2009) Association of a nsSNP in ADAMTS14 to some osteoarthritis phenotypes. Osteoarthritis Cartilage 17: 321-327.
26. Broeren MG, Vermeij EA, Arntz OJ, Bennink MB, Sterken E, et al.(2014) The validation of disease-inducible promoter constructs

for gene therapy in rheumatoid arthritis and osteoarthritis in human THP-1 cells. Ann Rheum Dis.

27. Koboldt DC, Steinberg KM, Larson DE, Wilson RK, Mardis ER (2013) The next-generation sequencing revolution and its impact on genomics. Cell 155: 27-38.
28. Verbruggen G, WittoekR, Vander Cruyssen B, Elewaut D (2012) Tumour necrosis factor blockade for the treatment of erosive osteoarthritis of the interphalangeal finger joints: a double blind, randomised trial on structure modification. Ann Rheum Dis 71: 891-898.
29. Cohen SB, Proudman S, Kivitz AJ, Burch FX, Donohue JP, et al (2011) A randomized, double-blind study of AMG 108 (a fully human monoclonal antibody to IL-1R1) in patients with osteoarthritis of the knee. Arthritis Res Ther 13: R125.
30. Naito S, Takahashi T, Onoda J, Yamauchi A, Kawai T, et al (2012) Development of a neutralizing antibody specific for the active form of matrix metalloproteinase-13. Biochemistry 51: 8877-8884.

CHAPTER 17

DETECTION OF MITOCHONDRIAL MUTATIONS IN CANCER BY NEXT GENERATION SEQUENCING

Fei Ye, David C, Samuels and Yan Guo

Department of Cancer Biology, Vanderbilt University, USA

ABSTRACT

Mitochondria play a central role in the regulation of cellular function, metabolism, and cell death in many types of cell including cancer. Little attention has been paid to the potential mutations that can affect mitochondrial function in cancer, other than specific mutations in genes that are encoded in the nuclear DNA and so once mutated, all mitochondria in a cell will be affected. However, both somatic and germline mutations in mitochondrial genes are common and have been suggested to contribute to the development of cancer. Recent advances in high-throughput technologies have allowed for the rapid and accurate detection of mitochondrial DNA (mtDNA) mutations, polymorphisms, or copy number variations in a variety of tissues and bodily fluids, and can be used as a tool for early detection and treatment of cancer. In addition to deep-sequencing specifically targeted at mtDNA, mtDNA sequences can also be extracted from exome (or whole genome) sequencing data.

Keywords

Sequencing, Mitochondria, Phosphorylation, Neurological diseases

MITOCHONDRIAL AND CANCER

Mitochondria generate the majority of cellular energy through oxidative phosphorylation, which produces ATP. Mitochondrial dysfunctions are an important cause of many neurological diseases [1] and drug toxicities [2,3] and may contribute to carcinogenesis and tumor progression [4,5] yet their role in tumorigenesis remains largely unknown. Early in the 1930s, Warburg has already discovered that cancer cells rely heavily on glycolysis to meet their metabolic demands (i.e., "aerobic-glycolysis") [6], and several hypothetical mechanisms have been proposed to explain this phenomenon [7]. In 1998, somatic homoplasmic mutations in tumor mtDNA were found that were not present in matched control tissues [8]. Since then, various studies have shown that mutations in mtDNA can contribute to cancer etiology [9,10] and mtDNA mutations are associated with various types of cancer, including breast cancer [11], prostate cancer [12,13], head and neck cancer [14], colon cancer [15], and bladder cancer [16,17]. It is likely that hereditary mitochondrial DNA variations may predispose individuals to cancer, while somatic mutations may possibly affect tumor progression. The functional consequences of mitochondrial mutations largely result in metabolic alteration [18-21] and changes in the protein composition of the mitochondrial inner membrane.

In breast cancer mutations in mitochondrial DNA have been linked to increased breast cancer risk through a deficiency in Electron Transport Chain (ETC) function and altered Reactive Oxygen Species (ROS) levels [22]. Altered expression levels of the Oxidative Phosphorylation System (OXPHOS) subunits or mitochondrial structural injury can impair ATP content and occur in breast- Infiltrating Ductal Carcinoma (IDC) [23]. Variations in the mtDNA copy number may be the overall result of gene (hereditary) and environmental interactions (oxidative stress) caused by potential exogenous cancer risk factors such as age, hormones, diet and environmental oxidants/antioxidants, and reaction to oxidative damage [24-26]. While it is difficult to prove the involvement of mtDNA mutations in triggering oncogenesis, there is increasing evidence that mutated mtDNA is a marker of poor survival in cancer prognosis. The first suggestions that mtDNA mutations may play a role in metastasis came from the comparison of frequency of somatic mtDNA mutations in Non-Small-Cell Lung Cancer (NSCLC) at different stages of tumor formation, which show a significantly decreased survival among advanced NSCLC patients harboring mtDNA mutations. However, very few studies have investigated the role of mtDNA copy number in breast cancer patients and those studies have generated somewhat conflicting results. Researchers have found that both increased mtDNA copy number from whole blood DNA [27] and reduced mtDNA copy number in tissue may increase breast cancer risk [28-30]. This understudied area needs further exploration.

MTDNA SEQUENCING

The number of mitochondria in a cell varies greatly by organism and tissue type. Usually, there are about 100 mitochondria in each mammalian cell, and 2-10 copies of mtDNA can be harbored in each mitochondrion [31]. Thus, heteroplasmy is a unique feature often observed with mtDNA mutations. A mtDNA mutation is heteroplasmic, if it is a mixture of normal and mutant mtDNA copies within a cell [32,33]. It has been found that heteroplasmies affecting the entire mitochondrial genome are common in normal individuals and moreover, the frequency of heteroplasmic variants varied considerably between different tissues in the same individual, although some of these mutations might have been germline and not somatic [34]. Researchers have found that heteroplasmy could promote tumor growth and that it is closely associated with aging [35-37]. Some studies even suggested that an mtDNA mutation does not need to reach homoplasmy, i.e. all copies of mtDNA within a cell are mutated, to promote tumor growth [24,25] Although this claim still lack definitive proof.

In many earlier studies on mtDNA mutations, data were obtained by Sanger sequencing, which cannot detect low abundant heteroplasmies. For more than ten years, a growing number of articles have described somatic mutations of the mitochondrial genome (mtDNA) in human tumors, identified by comparing tumor mtDNA and the mtDNA in adjacent normal tissue or blood based on Sanger sequencing. With the precise quantification achievable with deep sequencing, researchers can determine whether the cancer-specific somatic mutations are heteroplasmic rather than homoplasmic. Previously the sequencing accuracy was a concern in assessing the frequency of mutations in mitochondrial DNA, especially in tumor samples. By tagging both strands of duplex DNA, a recently proposed method called Duplex Sequencing may pave the way to explore the full power of high depth sequencing in answering fundamental questions, through its application to investigating the roles of mitochondria in cancer [38].

The short read sequencing market has been dominated by Illumina's sequencing platform. The most popular platform is HiSeq 2000/2500 which can output up to 200 million reads per sample. Because mitochondrion is very short, roughly 1 million 50 base pair long reads can achieve average depth of 3000x. Usually, multiplexing and barcoding technique will be used in association with mitochondria sequencing. Individual barcode (4 to 8 nucleotides) sequences are added to each sample so they can be distinguished after pooling. On a single lane in HiSeq 2500, up to 96 samples can be pooled together. Illumina's MiSeq platform offers longer read upto 300 bases pairs which allows for better de novo assembly and better detection of structural variants. But MiSeq's throughput is not comparable to HiSeq.

Other than performing high throughput sequencing targeted at mtDNA, mtDNA sequences can also be extracted from other sources of sequencing data. Because of the abundance of mtDNA, mtDNA sequence is one of the most common by product of sequencing data such as exome, whole genome and RNAseq [39]. In exome sequencing data, due to the lack of exome capture efficiency, a small portion of the reads will align to the mitochondrial genome, even when the mitochondrial genome is not the intended sequencing target. Although, the mitochondrial genome composed of almost all coding sequence which fits the definition of exome, however it is usually not included in the part of target DNA for the exome sequencing capture methods commonly used today. Recently, a study has shown that mitochondrial sequences can be extracted from exome sequencing data [40]. The average coverage of the mitochondrial genome from exome sequencing can be as high as several hundred (**Figure 1**), easily surpassing the average coverage of even the targeted exome regions [41]. The relatively high coverage of mtDNA is caused by the high copy number of mtDNA per cell. The advance of high-throughput sequencing technologies and the typically high coverage of a mtDNA sequence obtained provides a powerful tool for the study of mitochondrial DNA heteroplasmy in unprecedented detail [32,33,42-44]. This should be contrasted to techniques that specifically target the mitochondrial sequence, which can produce an average depth of tens of thousands of reads across the mitochondrial genome [34,42,45,46]. For mitochondria targeted sequencing, heteroplamsy as few as 0.1 percent are detectable [47]. For mitochondria sequences extracted from exome sequencing data, typically, heterplasmy of 1% is detectable.

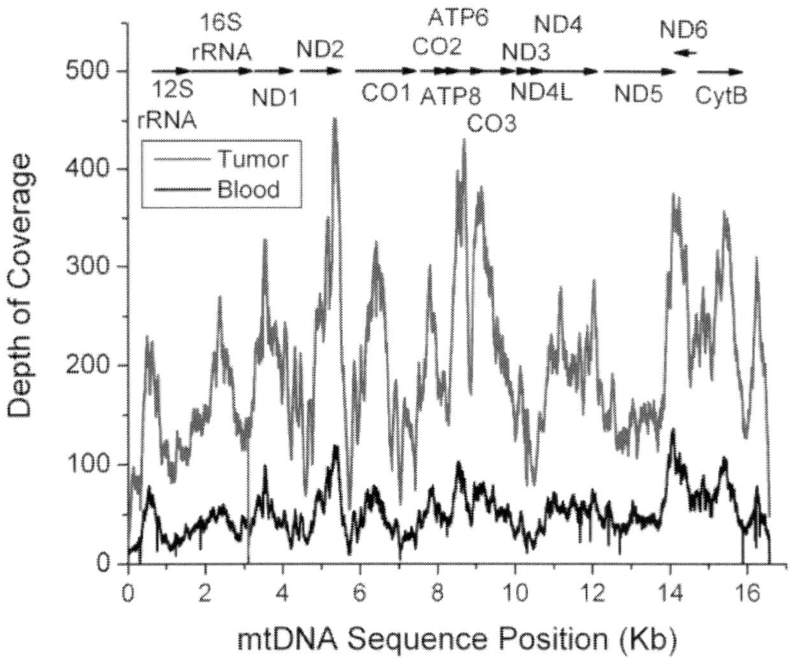

Figure 1: Example of mitochondria sequence depth in exome sequencing data. The red color denotes tumor and the black color denotes blood. The tumor tissue has more copies of mitochondria which resulted higher sequencing depth.

Multiple studies have started to infer mitochondria mutation information from exome sequencing data. The best example is The Cancer Genome Atlas (TCGA) project which contain thousands of exome sequencing data across 20 different types of cancers. Many study cohorts in TCGA have inferred mtDNA somatic mutation from exome sequencing data. Based on the recently published TCGA breast cancer marker paper [48], 325 mitochondrial DNA somatic mutations derived from off-target reads from the 776 exome sequencing data were reported. The concept of extracting mtDNA sequences from exome sequencing data can be easily applied to RNAseq and other type of targeted sequencing data.

mtDNA MUTATION DETECTION BY SEQUENCING

Sequencing alignment often suffers from ambiguity caused by homologous regions. This problem is amplified with mitochondria sequencing alignment due to the presence of nuclear copies of the mitochondrial genomes (nuMTS) [49,50]. As discussed previously, mitochondria sequences can be a byproduct of other sequencing data, nuclear DNA sequences can also be the byproduct of targeted mitochondria sequencing data. A read sequenced from nuMTS or mtDNA can cause ambiguity about whether it should align to the nuclear or the mitochondrial genome. There are three approaches dealing with the nuMTS ambiguity ranging from tolerant to conservative and simple to complex.

The most tolerant and simplest approach is to align the raw sequence reads against the mitochondrial reference genome directly and then throw out the non-aligned reads. This approach is simple because it ignores the nuMTS in the alignment and it is tolerant because it allows more reads to be aligned. However, this approach also allows nuMTS reads to be aligned to mitochondria genome which nuMTS introduce false heteroplasmic variability in the reported mtDNA sequence. The most conservative and complex approach is to first aligning reads against the nuclear genome only and then align the non-aligned reads to the mitochondrial genome. This approach is most conservative because it will allow mtDNA reads align to nuMTS which resulting less reads aligned to mitochondria genome and it is the most complex because it requires two alignment steps at the cost of additional alignment time and data storage.

A compromised approach is to align the raw reads against both the nuclear and mitochondrial genomes simultaneously. It is possible that a read can have two best mapping options, one at nuMTS and one at mitochondria genome. When such scenario happens, we are relied on the aligner used to make the correct decision. Most aligner such as BWA [51] and Bowtie [52] have the options to let the user to choose if this read is counted as 1 mapped randomly in nuMTS and mitochondria genome or counted as a fraction in all of the best mapping matches. The compromised approach has the best balanced between time consumption and misalignment rate. Mito Seek [53] software developed to extract mitochondria sequences from exome sequencing data implemented this approach.

Mitochondrial DNA copy number is highly variable and has been suggested to be associated with many diseases including cancer [27-30]. The traditional methods for assessing mtDNA copy number usually involve PCR [54] technology. A recently a novel method for evaluating mtDNA copy number was developed based on both sequencing-based assay of mitochondria NDA copy number and techniques developed for RNA expression profiling [55].

Mitochondria copy number can also be estimated from exome sequencing data. For example, it has been shown that the *fraction of captured mitochondrial sequences in exome sequencing data is proportionalto the relative abundance of the corresponding mitochondrialgenome in the original total DNA extract* [22,38,41]. The mtDNA copy extracted from exome sequencing data can be useful when studying tumor samples for conducting association tests with phenotypes such as the tumor stage and metastasis stage. MitoSeek can be used to compute relative mtDNA copy number from exome sequencing data.

CONCLUSIONS AND FUTURE PROSPECTS

Mutations of the mtDNA, heteroplasmic in most cases, were detected in different types of cancers, often paired with bodily fluids samples. Studies have also shown mtDNA mutations associated with cancer recurrence [56], genotoxic damage [57], and poor survival in late-stage cancer patients [58,59]. However, despite their high frequency in most solid human tumor types investigated so far, a causal role for a significant portion of mtDNA mutations in cancer cannot be claimed. Although some mtDNA mutations may enhance tumor growth, the enrichment of mutated mtDNA in solid human tumors can be explained by random drift, without need of any selective advantage [60]. More than 30 years after Sanger and colleagues first sequenced human mitochondrial genome, with the power of high depth sequencing technology, we are now able to study the role of mtDNA mutations in cancer to an extent not possible before, especially for off-target mtDNA heteroplasmy. Extending and improving the use of high depth sequencing is thus connected with great expectations to reject or retain current hypotheses proposed to explain the Warburg effect [7]. Furthermore, by assessing mitochondrial DNA content, exome sequencing data has also been used to diagnose certain mitochondrial disorders [61]. Additionally, high depth mtDNA sequencing may bring valuable insights to investigate if mtDNA mutations are associated with RNA-seq gene expression profiles in cellular pathways that regulate mitochondrial DNA synthesis and that are either directly or indirectly targeted by the tested anticancer drugs. MtDNA-dependent gene expression profiles can be identified to determine whether they reflect possible acquired mutations in the nuclear genome and to further investigate possible relationships between mitochondrial activity and gene expression changes, in both normal and cancer cells.

ACKNOWLEDGEMENTS

We would like to thank Margot Bjoring for her editorial support. Yan Guo, and Fei Ye were supported by grant CCSG (P30 CA068485)

REFERENCES

1. Fernandez-Vizarra E, Bugiani M, Goffrini P, Carrara F, Farina L, et al. (2007) Impaired complex III assembly associated with BCS1L gene mutations in isolated mitochondrial encephalopathy. Hum Mol Genet 16: 1241-1252.
2. Lemasters JJ, Qian T, Bradham CA, Brenner DA, Cascio WE, et al. (1999) Mitochondrial dysfunction in the pathogenesis of necrotic and apoptotic cell death. J Bioenerg Biomembr 31: 305-319.
3. Wallace KB, Starkov AA (2000) mitochondrial targets of drug toxicity. Annu Rev Pharmacol Toxicol 40: 353-388.
4. Modica-Napolitano JS, Singh KK (2004) Mitochondrial dysfunction in cancer. Mitochondrion 4: 755-762.
5. Chen EI (2012) Mitochondrial dysfunction and cancer metastasis. J Bioenerg Biomembr 44: 619-622.
6. Kaiser-Wilhelm-Institutfu\r Biologie B-D, Warburg OH (1930) The metabolism of tumours : investigations from the Kaiser Wilhelm

Institute for Biology, Berlin-Dahlem.

7. Kroemer G (2006) Mitochondria in cancer. Oncogene 25: 4630-4632.
8. Polyak K, Li Y, Zhu H, Lengauer C, Willson JK, et al. (1998) Somatic mutations of the mitochondrial genome in human colorectal tumours. Nat Genet 20: 291-293.
9. Baysal BE, Ferrell RE, Willett-Brozick JE, Lawrence EC, Myssiorek D, et al. (2000) Mutations in SDHD, a mitochondrial complex II gene, in hereditary paraganglioma. Science 287: 848-851.
10. Vanharanta S, Buchta M, McWhinney SR (2004) Early-onset renal cell carcinoma as a novel extraparaganglial component of SDHB-associated heritable paraganglioma. American journal of human genetics 74:153-159.
11. Canter JA, Kallianpur AR, Parl FF, Millikan RC (2005) Mitochondrial DNA G10398A polymorphism and invasive breast cancer in African-American women. Cancer Res 65: 8028-8033.
12. Herrmann PC, Gillespie JW, Charboneau L, Bichsel VE, Paweletz CP, et al. (2003) Mitochondrial proteome: altered cytochrome c oxidase subunit levels in prostate cancer. Proteomics 3: 1801-1810.
13. Petrosillo G, Di Venosa N, Ruggiero FM, Pistolese M, D'Agostino D, et al. (2005) Mitochondrial dysfunction associated with cardiac ischemia/reperfusion can be attenuated by oxygen tension control. Role of oxygen-free radicals and cardiolipin. Biochim Biophys Acta 1710: 78-86.
14. Sun W, Zhou S, Chang SS, McFate T, Verma A, Califano JA et al. (2009) Mitochondrial mutations contribute to HIF1alpha accumulation via increased reactive oxygen species and up-regulated pyruvate dehydrogenease kinase 2 in head and neck squamous cell carcinoma. Clin Cancer Res 15:476-484.
15. Ericson NG, Kulawiec M, Vermulst M, Sheahan K, O'Sullivan J, et al. (2012) Decreased mitochondrial DNA mutagenesis in human colorectal cancer. PLoS Genet 8: e1002689.
16. Dasgupta S, Hoque MO, Upadhyay S, Sidransky D (2008) Mitochondrial cytochrome B gene mutation promotes tumor growth in bladder cancer. Cancer Res 68: 700-706.

17. Fliss MS, Usadel H, Caballero OL, Wu L, Buta MR, et al. (2000) Facile detection of mitochondrial DNA mutations in tumors and bodily fluids. Science 287: 2017-2019.

18. Lundholm K, Edström S, Karlberg I, Ekman L, Scherstén T (1982) Glucose turnover, gluconeogenesis from glycerol, and estimation of net glucose cycling in cancer patients. Cancer 50: 1142-1150.

19. Mazurek S, Boschek CB, Eigenbrodt E (1997) The role of phosphometabolites in cell proliferation, energy metabolism, and tumor therapy. J BioenergBiomembr 29: 315-330.

20. Ockner RK, Kaikaus RM, Bass NM (1993) Fatty-acid metabolism and the pathogenesis of hepatocellular carcinoma: review and hypothesis. Hepatology 18: 669-676.

21. Fischer CP, Bode BP, Souba WW (1998) Adaptive alterations in cellular metabolism with malignant transformation. Ann Surg 227: 627-634.

22. Bai RK, Leal SM, Covarrubias D, Liu A, Wong LJ (2007) Mitochondrial genetic background modifies breast cancer risk. Cancer Res 67: 4687-4694.

23. Putignani L, Raffa S, Pescosolido R, Rizza T, Del Chierico F, et al. (2012) Preliminary evidences on mitochondrial injury and impaired oxidative metabolism in breast cancer. Mitochondrion 12: 363-369.

24. Lee HC, Lu CY, Fahn HJ, Wei YH (1998) Aging- and smoking-associated alteration in the relative content of mitochondrial DNA in human lung. FEBS Lett 441: 292-296.

25. Renis M, Cantatore P, LoguercioPolosa P, Fracasso F, Gadaleta MN (1989) Content of mitochondrial DNA and of three mitochondrial RNAs in developing and adult rat cerebellum. J Neurochem 52: 750-754.

26. Verma M, Naviaux RK, Tanaka M, Kumar D, Franceschi C, et al. (2007) Meeting report: mitochondrial DNA and cancer epidemiology. Cancer Res 67: 437-439.

27. Shen J, Platek M, Mahasneh A, Ambrosone CB, Zhao H (2010) Mitochondrial copy number and risk of breast cancer: a pilot study. Mitochondrion 10: 62-68.

28. Yu M, Zhou Y, Shi Y, Ning L, Yang Y, et al. (2007) Reduced

mitochondrial DNA copy number is correlated with tumor progression and prognosis in Chinese breast cancer patients. IUBMB Life 59: 450-457.

29. Tseng LM, Yin PH, Chi CW, Hsu CY, Wu CW, et al. (2006) Mitochondrial DNA mutations and mitochondrial DNA depletion in breast cancer. Genes Chromosomes Cancer 45: 629-638.
30. Bai RK, Chang J, Yeh KT, Lou MA, Lu JF, et al. (2011) Mitochondrial DNA content varies with pathological characteristics of breast cancer. J Oncol 2011: 496189.
31. Robin ED, Wong R (1988) Mitochondrial DNA molecules and virtual number of mitochondria per cell in mammalian cells. J Cell Physiol 136: 507-513.
32. Ng SB, Buckingham KJ, Lee C, Bigham AW, Tabor HK, et al. (2010) Exome sequencing identifies the cause of a mendelian disorder. Nat Genet 42: 30-35.
33. 1000 Genomes Project Consortium, Abecasis GR, Altshuler D, Auton A, Brooks LD, et al. (2010) A map of human genome variation from population-scale sequencing. Nature 467: 1061-1073.
34. He Y, Wu J, Dressman DC, Iacobuzio-Donahue C, Markowitz SD, et al. (2010) Heteroplasmic mitochondrial DNA mutations in normal and tumour cells. Nature 464: 610-614.
35. Sondheimer N, Glatz CE, Tirone JE, Deardorff MA, Krieger AM, et al. (2011) Neutral mitochondrial heteroplasmy and the influence of aging. Hum Mol Genet 20: 1653-1659.
36. Smigrodzki RM, Khan SM (2005) Mitochondrial microheteroplasmy and a theory of aging and age-related disease. Rejuvenation Res 8: 172-198.
37. Kann LM, Rosenblum EB, Rand DM (1998) Aging, mating, and the evolution of mtDNAheteroplasmy in Drosophila melanogaster. ProcNatlAcadSci U S A 95: 2372-2377.
38. Schmitt MW, Kennedy SR, Salk JJ, Fox EJ, Hiatt JB, et al. (2012) Detection of ultra-rare mutations by next-generation sequencing. ProcNatlAcadSci U S A 109: 14508-14513.
39. Samuels DC, Han L, Li J, Quanghu S, Clark TA, et al. (2013) Finding the lost treasures in exome sequencing data. Trends Genet

29: 593-599.

40. Larman TC, DePalma SR, Hadjipanayis AG; Cancer Genome Atlas Research Network, Protopopov A, Zhang J, et al. (2012) Spectrum of somatic mitochondrial mutations in five cancers. ProcNatlAcadSci U S A 109: 14087-14091.
41. Picardi E, Pesole G (2012) Mitochondrial genomes gleaned from human whole-exome sequencing. Nat Methods 9: 523-524.
42. Tang S, Huang T (2010) Characterization of mitochondrial DNA heteroplasmy using a parallel sequencing system. Biotechniques 48: 287-296.
43. Guo Y CQ, Samuels DC, Ye F, Long J, Li CI, Winther JF, et al. (2012) The use of next generation sequencing technology to study the effect of radiation therapy on mitochondrial DNA mutation. Mutation Research - Genetic Toxicology and Environmental Mutagenesis.
44. Goto H, Dickins B, Afgan E, Paul IM, Taylor J, et al. (2011) Dynamics of mitochondrial heteroplasmy in three families investigated via a repeatable re-sequencing study. Genome Biol 12: R59.
45. Guo Y, Cai Q, Samuels DC, Ye F, Long J, et al. (2012) The use of next generation sequencing technology to study the effect of radiation therapy on mitochondrial DNA mutation. Mutat Res 744: 154-160.
46. Ameur A, Stewart JB, Freyer C, Hagström E, Ingman M, et al. (2011) Ultra-deep sequencing of mouse mitochondrial DNA: mutational patterns and their origins. PLoS Genet 7: e1002028.
47. Guo Y, Li CI2, Sheng Q3, Winther JF4, Cai Q5, et al. (2013) Very low-level heteroplasmymtDNA variations are inherited in humans. J Genet Genomics 40: 607-615.
48. Cancer Genome Atlas Network (2012) Comprehensive molecular portraits of human breast tumours. Nature 490: 61-70.
49. Hazkani-Covo E, Zeller RM, Martin W (2010) Molecular poltergeists: mitochondrial DNA copies (numts) in sequenced nuclear genomes. PLoS Genet 6: e1000834.
50. Li M, Schroeder R, Ko A, Stoneking M (2012) Fidelity of capture-enrichment for mtDNA genome sequencing: influence of NUMTs.

Nucleic Acids Res 40: e137.

51. Li H, Handsaker B, Wysoker A, Fennell T, Ruan J, et al. (2009) The Sequence Alignment/Map format and SAMtools. Bioinformatics 25: 2078-2079.

52. Langmead B, Salzberg SL (2012) Fast gapped-read alignment with Bowtie 2. Nat Methods 9: 357-359.

53. Guo Y, Li J, Li CI, Shyr Y, Samuels DC (2013) MitoSeek: extracting mitochondria information and performing high-throughput mitochondria sequencing analysis. Bioinformatics 29: 1210-1211.

54. Bhat HK, Epelboym I (2004) Quantitative analysis of total mitochondrial DNA: competitive polymerase chain reaction versus real-time polymerase chain reaction. J BiochemMolToxicol 18: 180-186.

55. Castle JC, Biery M, Bouzek H, Xie T, Chen R, et al. (2010) DNA copy number, including telomeres and mitochondria, assayed using next-generation sequencing. BMC Genomics 11: 244.

56. Fliss MS, Usadel H, Caballero OL, Wu L, Buta MR, et al. (2000) Facile detection of mitochondrial DNA mutations in tumors and bodily fluids. Science 287: 2017-2019.

57. Wardell TM, Ferguson E, Chinnery PF (2003) Changes in the human mitochondrial genome after treatment of malignant disease. Mutation Research/Fundamental and Molecular Mechanisms of Mutagenesis 525:19-27.

58. Matsuyama W, Nakagawa M, Wakimoto J, Hirotsu Y, Kawabata M, et al. (2003) Mitochondrial DNA mutation correlates with stage progression and prognosis in non-small cell lung cancer. Hum Mutat 21: 441-443.

59. Lièvre A, Chapusot C, Bouvier AM, Zinzindohoué F, Piard F, et al. (2005) Clinical value of mitochondrial mutations in colorectal cancer. J ClinOncol 23: 3517-3525.

60. Kirches E (2009) Mitochondrial and nuclear genes of mitochondrial components in cancer. Curr Genomics 10: 281-293.

61. Dinwiddie DL, Smith LD, Miller NA, Atherton AM, Farrow EG, et al. (2013) Diagnosis of mitochondrial disorders by concomitant next-generation sequencing of the exome and mitochondrial genome. Genomics 102: 148-156.

CHAPTER 18

NEXT GENERATION SEQUENCING: ADVANCES IN CHARACTERIZING THE METHYLOME

Kristen H. Taylor [1], Huidong Shi [2] and Charles W. Caldwell [1]

[1]University of Missouri-Columbia School of Medicine, Ellis Fischel Cancer Center, Columbia, MO 65212, USA

[2]Medical College of Georgia, Augusta, GA 30912, USA

ABSTRACT

Epigenetic modifications play an important role in lymphoid malignancies. This has been evidenced by the large body of work published using microarray technologies to generate methylation profiles for numerous types and subtypes of lymphoma and leukemia. These studies have shown the importance of defining the epigenome so that we can better understand the biology of lymphoma. Recent advances in DNA sequencing technology have transformed the landscape of epigenomic analysis as we now have the ability to characterize the genome-wide distribution of chromatin modifications and DNA methylation using next-generation sequencing. To take full advantage of the throughput of next-generation sequencing, there are many methodologies that have been developed and many more that are currently being developed. Choosing the appropriate methodology is fundamental to the outcome of next-generation sequencing studies. In this review, published technologies and methodologies applicable to studying the methylome are presented. In addition, progress towards defining the methylome in lymphoma is discussed and prospective directions that have been made possible as a result of next-generation sequencing technology. Finally, methodologies are introduced that have not yet been published but that are being explored in the pursuit of defining the lymphoma methylome.

Keywords: Lymphoma, leukemia, next-generation sequencing, methylation, epigenome

INTRODUCTION

The human genome is comprised of three billion base pairs of DNA, of which ~30 million are CpG dinucleotides, the main sites of DNA methylation. Methylation of cytosine residues at CpG dinucleotides is a prevalent epigenetic modification in mammalian genomes which is known to have profound effects on gene expression [1,2,23]. This epigenetic event occurs globally in the normal genome and is estimated to affect between 70 and 80% of all CpG dinucleotides in human cells [4,5]. These dinucleotides are not uniformly distributed across the genome but occur in clusters such as large repetitive sequences or in CG-rich DNA stretches known as CpG islands (CGIs). The majority of the CGI which are found in intragenic regions, including repetitive sequences such as satellite sequences and centromeric repeats, contain methylated CpG dinucleotides while CGIs which are found preferentially in the promoter regions of genes typically contain unmethylated CpG dinucleotides [6]. Some exceptions to this rule include those CGIs located on the inactive X chromosome in females [7] and those associated with imprinted genes (genes for which only the paternally- or maternally-inherited allele is expressed) which are methylated in the normal state [8,9]. Contrary to normal cells, a number of studies examining DNA methylation in human cancers have shown global hypomethylation and concomitant hypermethylation in the regulatory regions of specific genes [10,11], although this concept has recently been challenged [12]. Methylation of DNA in humans is generally thought to take place via some combination of three DNA methyltransferase enzymes (DNMT1, DNMT3A, and DNMT3B). Whereas DNMT1 is responsible for the maintenance of normal methylation patterns, DNMT3A and 3B are primarily associated with *de novo* methylation. Large-scale studies of DNA methylation have shown that aberrant methylation is a phenomenon present in virtually all types of tumors but it occurs in a generally non-random manner that differs between tumor types. Exactly how the methylation is directed and controlled is still a largely unanswered question. Polycomb group (PcG) proteins are likely to be key in marking certain genes for DNA methylation and in altering histone methylation leading to the formation of repressive chromatin, but the mechanisms by which this occurs are not completely elucidated [13,14,15,16,17,18]. A class of non-coding RNAs, the microRNAs, are also likely modulators of DNA methylation. It has been suggested that miR-143 controls DNMT3A in colorectal cancer [19] and that the miR-29 family regulates DNMT3A and 3B in lung cancer [20]. Furthermore, miR-29b has recently been shown to target all three DNA methyltransferase genes either directly or indirectly in acute myeloid leukemia [21].

A number of methods have been used to examine DNA methylation at whole- or sub-genome scale, and over time these methods have progressed from the ability to characterize the methylation profiles of candidate genes to the ability to develop genome-wide methylation profiles (Figure 1). These have been discussed in detail in an excellent review by Peter Laird [22]. Generally, these can be categorized into the preparative method and the analytical or detection systems employed. The preparative methods typically fall into three broad classes that include enzyme digestions, affinity enrichments, or treatment with sodium bisulfite while the detection systems can be described as locus-specific, gel-based, array-based, or as discussed below, next-generation sequencing (NGS) based methods. Although none of these earlier methods is perfect, each has allowed contributions of valuable epigenetic data to the scientific community from a large number of laboratories. In the advent of next generation sequencing, we now have the ability to detect aberrant epigenetic events in cancer at high resolution. This review describes the technologies and methods currently being used to develop high resolution methylation profiles and the progress that has been made using NGS to characterize the methylome in lymphoma.

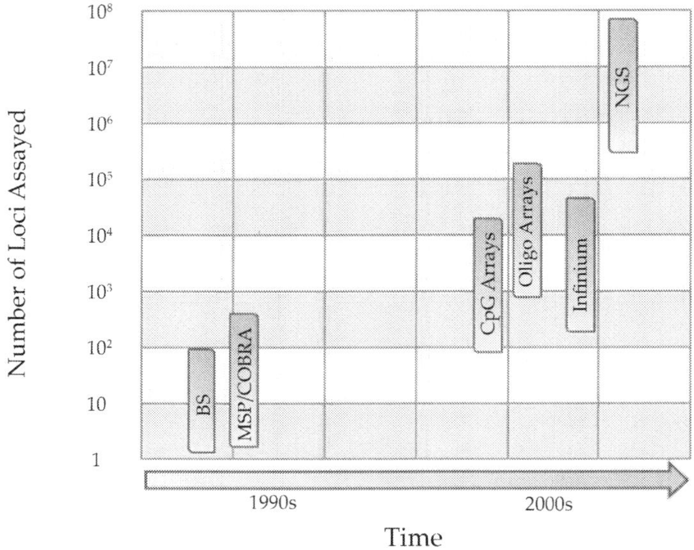

Figure 1. Time *versus* number of loci assayed. The number of loci assayed has increased over time with the development of new technologies. Bisulfite sequencing (BS), methylation specific PCR (MSP) and combined bisulfite restriction analysis (COBRA) were introduced in the 1990s. These methods provided methylation data for up to a few hundred genes in a single experiment. Methylation arrays were introduced in the 2000s and provide methylation data for up to 100s of thousands of loci. (N.B. The latest oligonucleotide array available from Nimblegen allows for the interrogation of more than two million loci). Next-generation sequencing (NGS) was introduced in 2007 and provides single base pair resolution for billions of nucleotides.

CUTTING-EDGE TECHNOLOGICAL ADVANCES

In a relatively short time period, since its inception in 2007, the power of NGS technology has transformed the science of genomic research. Currently, five commercial NGS platforms are available; however, Roche/454 (Branford, CT), Illumina/Solexa (San Diego, CA), and Applied Biosystems/SOLiD™ (Carlsbad, CA), dominate the sequencing market shares. Additional platforms such as Pacific Biosciences (Menlo Park, CA), are on the horizon. A feature that all NGS platforms share is the parallel sequencing of clonally amplified or single DNA molecules that are immobilized on an array or flow cell. This feature fundamentally sets NGS apart from conventional capillary-based sequencing in that millions of sequence reads can be generated in parallel rather than 96 at a time. The significant increase in throughput has lead to a dramatic reduction in sequencing cost and a massive accumulation of genomic sequences. The Human Genome Project which was completed in 2003 by Sanger sequencing took 13 years to complete and cost approximately $2.7 trillion dollars. Now, in 2010, a human genome can be sequenced in eight days with 30x coverage for approximately $10,000 dollars using the latest HiSeq2000 sequencer from Illumina. The NGS platforms have facilitated many novel sequencing applications, particularly in the area of epigenomic research. For example, in 2009, the first human epigenome was sequenced using NGS [23]. Although all NGS platforms share some common features, they also differ significantly. (See [24] for an in depth review of the sequencing technologies and a comparison of current specifications). These differences will affect the applications of each platform and the challenges of bioinformatic analysis as described below.

Roche (454) Genome Sequencer FLX

The 454 Genome Sequencer FLX (454 GS FLX) was the first commercially available NGS to combine single-molecule emulsion PCR with massively parallel pyrosequencing. The sequencing chemistry is based on the principle of pyrosequencing in which a pyrophosphate molecule is released upon nucleotide incorporation by DNA polymerase for luminescence detection. The clonal amplification of sequencing templates bound on agarose beads is achieved by emulsion PCR which uses a vigorously mixed oil and aqueous mixture to create water-in-oil microvesicles. These microvesicles serve as micro-bioreactors for the PCR reaction. Under limiting-dilution conditions only one bead will be present in each water-in-oil vesicle. Each agarose bead surface contains several million copies of the same single stranded sequencing template, while millions of such beads can be generated in a matter of hours [25]. The microbeads are enriched and deposited into 3.4 million picoliter-scale sequencing reaction wells and combined with sequencing chemistry (See http://www.454.com for an explanation of the workflow and technology). Each picotiter plate can be divided into 16 portions allowing for the simultaneous sequencing of 2, 4, 8 or 16 samples. Over one hundred bar-coded sequencing adaptors are available and therefore up to a thousand samples can be multiplexed on one picotiter plate. The biggest advantage of 454's sequencing technology is the read length which is on average 450 bp using the newest "titanium" version compared to 50 to 125 bp read lengths generated by the other two major platforms. The longer reads facilitate *de novo* assembly for certain sequencing applications. About 1GB of data from one million sequencing reads can be generated in a single seven hour run on the 454 GS FLX as opposed to a multiple day run on both the Illumina GA and ABI SOLiD™ platforms. However, the Illumina GA, and ABI SOLiD™ sequencers produce approximately 100 million reads per run. Therefore, the sequencing cost per base is higher with the 454 GS FLX. The 454 GS FLX was also the first NGS platform to be used for epigenetic studies. The 454 website currently lists 23 publications in the field of epigenetics and of these four used bisulfite sequencing to determine the precise location of CpG methylation in cancer [26,27,28,29]. One major limitation of the 454 technology is that it is difficult to correctly call homopolymers greater than 3-4 bases in length. This limitation is exasperated in bisulfite sequencing because in many GC rich regions, a long stretch of poly-A or poly-T will appear due to the conversion of unmethylated C to T. However, since the bisulfite sequence reads are not assembled but merely aligned to the reference genome sequence, modifications to the mapping algorithms can solve this problem. In addition, the longer sequencing reads produced by this technology greatly enhances the opportunity for accurate alignment of the bisulfite converted sequences to the reference genome. Approximately 70% of bisulfite sequencing reads generated by the 454 GS FLX can be mapped accurately to the reference genome while there is significant variation reported in the mapping rates using the Illumina GA and SOLiD™ platforms due to the shorter read lengths produced by these technologies [30,31,32].

Illumina Genome Analyzer

The Illumina GA was the first short read NGS platform based on the concept of "sequence-by-synthesis" (SBS). It produces sequence reads of 36 to 125 bp from tens of millions of clonally amplified DNA fragments on a flow cell. The Illumina technology uses a novel PCR process known as bridge PCR to achieve clonal amplification of sequencing templates on the surface of a glass flow cell comprised of eight separate lanes. Each lane has covalently attached oligonucleotides that are complimentary to specific adaptors that are ligated to the library fragments. Under limiting-dilution conditions, adaptor-modified single-stranded template DNA is hybridized to the anchor oligonucleotides on the flow cell followed by subsequent cyclic PCR to convert a single DNA molecule into a clonally amplified cluster consisting of approximately 1000 molecules. The Illumina sequencing chemistry uses four different fluorescent dye labeled nucleotides and resembles traditional Sanger sequencing. However, these nucleotides are modified at the 3'-OH group to ensure that only a single nucleotide is incorporated in each sequencing cycle. Each sequencing cycle is followed by an imaging step to capture the incorporated nucleotide at each cluster. The fluorescent group is removed after imaging and the 3'-end is re-opened for the next base incorporation cycle. Up to 125 cycles can be repeated using existing protocols which generate up to 125 bp sequence reads due to the improvement of sequencing chemistry and image analysis software. A typical run yields approximately 70-100 million reads and takes three to six days depending on the read length desired. Because of its robust performance and the improvement of read length and accuracy over time, the Illumina GA occupies over 60% of the current market. Many different applications have been developed on this platform including RNA-seq, small RNA sequencing, ChIP-seq, and whole genome and targeted sequencing. The Illumina GA has been used for shotgun whole-genome bisulfite sequencing of human, mouse and plant genomes [23,33,34,35]. Also, a method known as targeted capture bisulfite sequencing was developed for the Illumina GA instrument [30,32,36]. In total there are more than 40 epigenetics publications using the Illumina GA instrument. Of these, 17 address DNA methylation and two address DNA methylation in cancer [32,37]. Many early bisulfite sequencing experiments produced less than 30% mappable reads compared to ~ 80-90% mappable reads for non bisulfite-treated sequencing using the same short read alignment software [30,32,36]. This discord in mappability is due in large part to the limitations of current mapping algorithms to align bisulfite converted sequences. Furthermore, because the mapping rate may vary significantly between samples, the comparability of data may be jeopardized.

Applied Biosystems (SOLiD™)

The SOLiD™ (Sequencing by Oligo Ligation and Detection) system was released in 2007 by Applied Biosystems and is based on a unique sequencing process called "sequencing by ligation". The approach was initially developed by George Church's group at Harvard and was termed polony sequencing [38]. Applied Biosystems refined the technology and developed a two color sequencing chemistry for this instrument. Similar to the 454 system, the SOLiD™ system also uses emulsion PCR to generate clonally amplified sequencing templates. However, much smaller magnetic beads (1 um) are used in the emulsion PCR process and the beads are covalently attached to the surface of glass slides. With the current SOLiD™ 4 system, over 700 million reads per run can be achieved for a single read library. Each run requires 6-8 days and produces 40-50 GB of data with a read length of 50 bp. In addition to genome sequencing, RNA-seq, small RNA sequencing, and ChIP-seq, bisulfite sequencing of a bacterial genome has recently been published using this technology [39]. The main advantages of this technology are increased throughput and increased sequencing accuracy compared to the other sequencing platforms.

Other Sequencing Technologies

Helicose is an alternative NGS platform which uses a similar principle to the three major platforms but does not rely on clonal amplification. It is the only platform to date that has truly achieved single molecule parallel sequencing. However, because of its late entry into the market, and a relatively high cost, sales have considerably lagged behind the three platforms described above. In addition, the Polonator has been commercialized by Dover systems and is based on the original polony sequencing methods developed by Church's group [38]. This system is relatively inexpensive but the throughput and read lengths are lower than the major commercial platforms; therefore, few laboratories are using this system. Complete Genomics, a company based in Mountain View, California, has also published a sequencing method [40]. Speed and cost have been Complete Genomics' key selling points. They claim that by June of 2010 the cost of materials will be approximately $1,000 per genome.

Third Generation Sequencing Technologies

In addition to the second generation sequencing platforms mentioned above, third generation sequencing platforms are beginning to emerge. For example, Pacific Biosciences uses a single-molecule technology with engineered DNA polymerase to capture and record the incorporation of fluorescently labeled nucleotides into growing complementary nucleic acid strands. While this technology produces fewer reads than 454's GS FLX, Illumina's GA and Applied Biosystem's SOLiD™, it is distinguished by its very long reads and ultra-fast real time sequencing. This transformative technology will enable a new paradigm in genomic analysis. Particularly interesting, the tempo and duration of the resulting fluorescent pulses yield a rich set of information about polymerase kinetics and allow direct detection of various forms of modified nucleotides, including 5-methylcytosine, 5-hydroxymethylcytosine, and N6-methyladenosine. Therefore, it is possible to detect methylated bases without the need for bisulfite conversion of genomic DNA. Unlike current bisulfite-sequencing techniques which are limited by short read lengths and by a reduction in genomic complexity, the Pacific Biosciences method will enable mapping of methylation patterns even within highly repetitive genomic regions.

METHYLOME METHODOLOGIES

Genome-wide methylation methodologies can be divided into those that divulge the precise location of methylated CpG dinucleotides (site specific) and those that disclose methylated regions of the genome. The site specific methods rely on bisulfite conversion of DNA while the regional methods rely on enzyme digestion or the affinity of either antibodies or proteins to methylated DNA. Sodium bisulfite converts unmethylated cytosines to uracils but has no effect on methylated cytosines. This conversion can then be exploited in a sequencing context because cytosines that were originally unmethylated will appear as thymines after amplification and those that were methylated will remain cytosines. The net effect of this conversion process is that the complexity of the genome is greatly reduced since consequently the genome now consists of three bases instead of four with the exception of methylated cytosines. Bias may be introduced in site specific methods due to incomplete bisulfite conversion and the efficiency of bisulfite PCR. These methods also require extensive bioinformatics for aligning the resulting sequences to the bisulfite converted genome and therefore may not be a suitable option for laboratories that do not have bioinformatic support.

Alternately, affinity based methods enrich for the methylated fraction of the genome by "pulling down" methylated DNA fragments. These methods are limited by the sensitivity and specificity of the antibody or binding protein employed. In general, these methods are more effective in enriching for regions of the genome with a high density of methylation and less efficient in CpG poor regions of the genome. Given that highly repetitive sequences are densely methylated, a considerable proportion of the sequences generated may be impossible to unambiguously align to the genome. In addition, digestion with methylation sensitive or methylation dependent enzymes enrich for the unmethylated or methylated regions of the genome respectively. When combined with NGS fragments that are 'missing' after digestion with a methylation sensitive enzyme represent the methylated regions of the genome while those that are present after digestion with a methylation dependent enzyme represent the methylated regions of the genome. Because the affinity and restriction enzyme methods confer only regional methylation information, additional experiments are needed to identify the precise location of CpG methylation within these regions.

Site Specific

Whole-genome bisulfite sequencing provides coverage at single base pair resolution and is therefore the most comprehensive of the site specific methodologies. In this method the DNA is sheared and bisulfite adaptors are ligated to the fragments. These adaptors can be synthesized with 5-methylcytosine so that the cytosines remain cytosines after bisulfite conversion or alternatively one can use modified amplification primers to include the bisulfite converted version of the adaptor. It is important to note that the sequence of the adaptors after bisulfite conversion must be complimentary to the seeding oligos. Therefore, it is imperative that the cytosines are preserved after conversion in order for the fragments to be clonally amplified on a flow cell. After the adaptors are added, the sample is bisulfite converted and the converted fragments are PCR amplified. These amplified fragments are then used in the library preparation for high throughput sequencing. In 2008, Cokus and colleagues published the complete methylome of Arabidopsis using this technique coupled with the Illumina GA sequencing platform [33]. This technology produced 3.8 billion high quality nucleotides providing 20X coverage of the 120 Mb Arabidopsis genome. Of these, 2.6 billion nucleotides mapped uniquely to the genome and covered 93% of all cytosines that could be theoretically covered. The size of the Arabidopsis genome makes it amenable to this type of study, however, the size of the human genome is approximately 150X larger than the Arabidopsis genome and therefore performing these experiments in human (or other mammals for that matter) is cost prohibitive at this time. Nonetheless, this technology was used by Laurent and colleagues to compare human cell types at three progressive stages of differentiation [41] and by Lister and colleagues to compare human embryonic stem cells and fetal fibroblasts [23]. To date, the only other group to use the shot-gun bisulfite sequencing approach constructed an *Escherichia coli* methylome using the SOLiD™ platform [39]. Whole-genome bisulfite sequencing provides unbiased coverage of the genome allowing for the interrogation of regions of the genome that are often missed using other methodologies.

While bisulfite treatment followed by shotgun sequencing remains a daunting task for large genomes, there are methods available that enrich CpG fragments which greatly reduces the depth of sequencing required for whole-genome coverage. Some of these methods are technically challenging and each has unique biases that are introduced. Reduced representation bisulfite sequencing, targeted capture and padlock probes are described in the following sections.

Reduced representation bisulfite sequencing (RRBS) enriches for CpG rich regions of the genome without the need for a known set of targets (*i.e.*, candidate genes, CpG islands, promoters, enhancers *etc.*). In this method DNA is fragmented with a restriction enzyme (or combination of enzymes), adaptors are ligated to size selected fragments then bisulfite treated and amplified to produce a library for NGS (Figure 2A). Enrichment using restriction enzymes is not truly genome-wide but includes a large portion of the genome and has the advantage of targeting non-repetitive CpG containing regions of the genome. The data generated includes regions of the genome that are in close proximity to a specific restriction enzyme's recognition sequence. This is helpful in regards to the bioinformatic analysis of these data because the genome is reduced and therefore less computational power is required to align the sequences. In choosing the appropriate enzyme to use to reduce the genome to be sequenced, one must consider which NGS platform will be used since each sequencing platform has unique properties. For example the Illumina and Applied Biosystems platforms produce short sequence reads and therefore the size of the restriction fragment produced should remain small while 454 can handle longer fragments and to reach its full potential restriction enzymes that result in larger fragments should be chosen in order to cover as much of the genome as possible.

Meissner and colleagues first published the concept of RRBS in 2005 [42]. After the restriction of mouse DNA with BglII (recognition sequence AGATCT), fragments between 500 and 600 bp were size selected and then adaptors were ligated to the size selected fragments. The fragments were then subjected to bisulfite treatment, amplified using bisulfite-modified adaptor sequences, cloned into a plasmid vector and then sequenced using conventional Sanger sequencing. To align the data, the generated sequences were searched against a reduced representation database of *in silico* size selected and bisulfite treated BglII fragments. This method was modified in 2009 by Meissner's group for amenability to the Illumina sequencing platform [43]. In this protocol MspI, a methylation insensitive enzyme, was used to digest mouse DNA. The use of this particular enzyme ensures that at least two data points will be produced for each restriction fragment since the MspI recognition sequence is CCGG. The resulting fragments were end repaired and then methylated adaptor sequences were ligated to the fragments. The adaptor-ligated fragments were size selected for 40-120 bp and 120-220 bp and two sequencing libraries were produced. The selected fragments were bisulfite converted and then amplified to produce two sequencing libraries instead of a single library to reduce size related amplification bias.

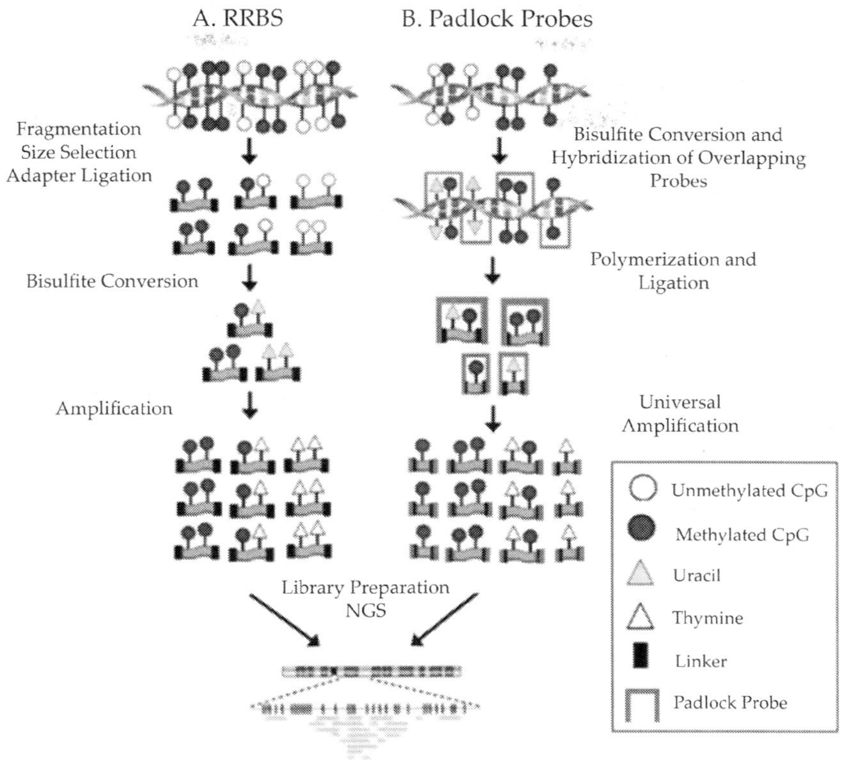

Figure 2. Site specific NGS methodologies. **A.** Reduced representation bisulfite sequencing (RRBS). **B.** Padlock probe assisted multiplex amplification. NGS: Next generation sequencing.

In addition to using single enzymes for RRBS, combinations of enzymes can also be used. Zeshnigk and colleagues used MseI (TTAA) and Tsp509I (AATT) to deplete AT rich regions of the genome in combination with NlaIII (CATG) and HpyCH4 (TGCA) [44]. After digestion, adaptors were ligated to blunt-ended and phosphorylated fragments. The adaptor-ligated fragments were treated with bisulfite, amplified and then size selected (440-900 bp) before sequencing using the 454 GS FLX. This study compared the methylation present in human female white blood cells with the methylation present in human sperm. Theoretically this combination of enzymes should produce data that covers 12,482 CGI (~45% of annotated CGI); however, only 21-22% of annotated CGI were covered and only 7.6% and 7.9% of the female white blood cell and sperm cell reads respectively mapped to a CGI. While this may be an effective enrichment strategy, these results suggest that this library preparation method is biased toward non-CG rich regions; however, it may be possible to increase the CGI coverage by obtaining more sequence reads.

Targeted capture microarrays comprised of probes that encompass regions of interest or targeted probe libraries within solution can be used to enrich for regions of the genome before sequencing. Targeted probes can be developed to capture bisulfite converted or non-converted sequences. An important consideration, if the bisulfite converted genome is the targeted genome, is that capture probes must be designed such that all permutations after bisulfite conversion are included. A clear advantage to this method is that it provides a reduced genome without the sequence constraint inherent in RRBS which utilizes restriction enzymes. In 2009 Hodges and colleagues published a method for targeting bisulfite converted DNA to enrich for regions of interest in the genome [32]. In this method, genomic DNA was sonicated and then methylated Illumina adaptors were ligated to the fragments. The fragments were then size selected (200-300 bp) and bisulfite converted. The converted templates were PCR amplified and captured on a custom Agilent 244 k array. The captured fragments were then eluted and sequenced using the Illumina GA platform. These authors sequenced a normal and a cancer cell line and report an enrichment of over 700 fold in the normal cell line and over 1300 fold in the cancer cell line. In both cell lines more than 90% of the target region was covered by at least 10 sequencing reads. However, only 6.43% of the total reads in the normal cell line and 11.98% of the total reads in the cancer cell line mapped to the targeted region. Therefore, a disadvantage of this method is that the capture is often imprecise and fairly inefficient. On the other hand, because an amplification step is required in this protocol, less DNA is required as starting material than is needed for arrays targeted at non-bisulfite converted DNA.

Padlock probe assisted multiplex amplification is performed in solution and has the capability of targeting more of the genome than the current array based capture platforms. Padlock probes are designed such that a region of interest is covered by overlapping probes on both the sense and antisense DNA strand (Figure 2B). Each probe consists of two target specific arms connected by a universal backbone sequence and hybridizes to the DNA in two places leaving the backbone sequence unbound. The gap between the two arms is polymerized and then ligated to form circular DNA. The backbone sequence is then used for universal PCR allowing for the amplification of tens of thousands of probes within a single reaction. Extensive bioinformatics is required for the design of the padlock probes because each probe must be normalized for melting temperature and length. Recently, two groups utilized this technology for methylation analysis. In each case, a library of padlock probes was annealed to bisulfite converted DNA. Next the 3' ends were extended and ligated to the 5' ends. The linear DNA was removed by exonucleases and then the circularized DNA was amplified with common primers. Deng and colleagues developed padlock probes to capture 175-225 bp regions including CpG islands, promoters and the transcription start site of genes involved in development or pluripotency [30]. In total 10,582 probes were required to cover 2.1 Mbp of the genome. However, since bisulfite conversion results in all unmethylated cytosines being converted to uracils, 30,000 probes were developed in which all C/T combinations were included for those probes that contained CpG sites. They observed a 10,000 fold difference in capture efficiencies between the most and the least efficient probes. Therefore, some probes required re-synthesis (to gain efficiency) and some probes required the addition of suppressor oligos (to reduce efficiency) which is the primary shortcoming of this method. Ball and colleagues focused on the ENCODE pilot project regions and avoided CpGs in the hybridizing arms so that only one probe was necessary per locus [36]. Their design targeted regions of the genome that are not associated with CGI and therefore provide data from regions not typically included in methylation studies.

Regional

Restriction-enzyme methods create a genome-wide methylation profile for regions of the genome that are encompassed by DNA fragments created after digestion with either a methylation sensitive or methylation dependent enzyme (or combination of enzymes). Ball and colleagues utilized the methylation sensitive enzyme HpaII to create a library of tag fragments that were then sequenced using the Illumina platform [45]. In this method, a library of tag fragments from all restriction fragments was created and then subjected to NGS. The methylation of a particular fragment was inferred based on the number of times it was observed. For example, fragments that were observed many times had low levels of methylation while those that were not observed had high levels of methylation. As a control, the authors also created an MspI library which has the same recognition sequence as HpaII but cuts either methylated or unmethylated CCGG sequences. After final analysis, the authors concluded that the control library was not necessary because the results from the HpaII library alone was highly correlated with the methylation present at individual sites. This method is better suited for distinguishing between highly and moderately methylated fragments as opposed to moderately and weakly methylated fragments because there is more noise associated with the weakly methylated (and therefore more highly digested) fragments. Concurrently, a second group published a modification of the HpaII tiny fragment enrichment ligation mediated PCR (HELP) protocol, HELP-seq [46]. For this method, the authors removed the adaptors from the LM-PCR products with MspI and then used the fragments to create sequencing libraries. The results of HELP-seq and HELP with microarray were compared and found to be strongly concordant. However, HELP-seq identified more hypomethylated loci and allowed the identification of shorter sequences that were not able to be represented on the microarray. In 2009, Wang and colleagues used a methylation dependent enzyme McrBC to identify methylation in the maize genome [47]. Cleavage with McrBC requires the presence of two methylated recognition sequences separated by 40-3000 base pairs. In this method, genomic DNA was digested with McrBC and then size selected fragments were used for library construction and then sequenced using the Illumina platform. Restriction based methods provide an excellent initial screen of the genome but are biased toward regions of the genome that contain restriction enzyme recognition sequences and require additional experiments to determine the precise location of methylated cytosines within the methylated fragments.

Affinity methods such as methylated DNA immunoprecipitation (MeDIP) and methylated CpG island recovery assay (MIRA) do not rely on the bisulfite modification of DNA [48,49]. MeDIP employs an antibody directed against 5-methylcytosine while MIRA (and modifications of MIRA) utilizes methyl-binding proteins (Figure 3). One major difference between the methods is that MeDIP requires DNA to be single-stranded in order to achieve efficient pull down which is sometimes difficult to achieve in regions of high CpG content and MIRA does not. Both MeDIP and MIRA can now be performed using commercial kits available from multiple companies. Briefly, for MeDIP, the DNA is sheared into 300-1 kb fragments, denatured and incubated with anti-5 methyl-cytosine.

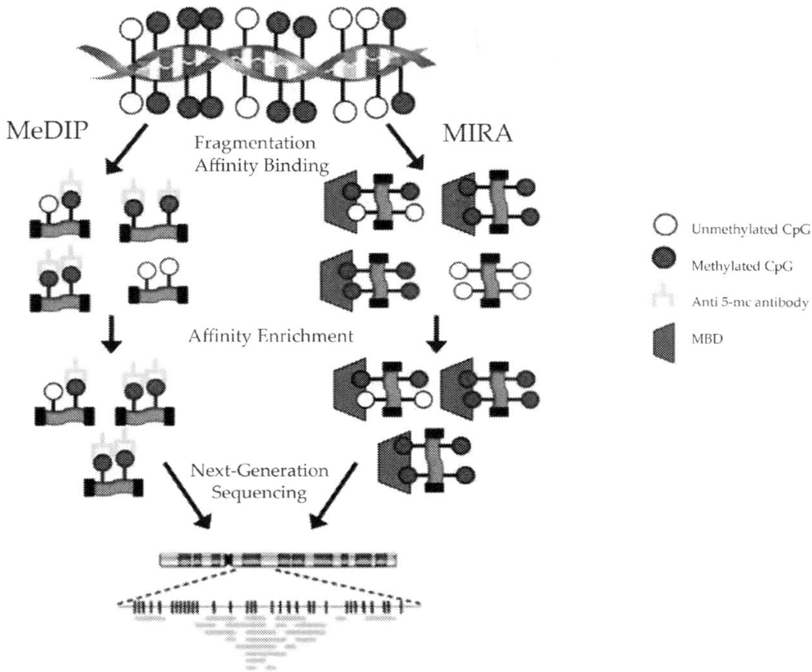

Figure 3. Affinity-based methodologies. MeDIP: Methylated DNA immunoprecipitation; MIRA: Methylated CGI recovery assay; MBD: Methyl-binding domain.

Magnetic or agarose (sepharose) beads are added and the methylated DNA is separated from the unmethylated DNA. Recently, Butcher and Beck published a modified version of this protocol amenable to high-throughput methylome analysis termed AutoMeDIP-seq [50]. In MIRA, a His-tagged MBD2b/MBD3L1 complex is incubated with DNA fragmented either by shearing or by digestion with MseI. The methylated protein-DNA complexes are captured using magnetic beads and the DNA is eluted while the protein complexes are degraded. A modification of this technique was recently used to characterize the methylome of three cancer cell lines and termed methyl binding domain (MBD) isolated genome sequencing [51]. Bioinformatic analysis of data generated after affinity enrichment is complicated by the need to take into consideration the density of CpG dinucleotides in a given region and by the inherent enrichment of repeat regions which can be difficult or even impossible to align to the genome. Overall, affinity methods provide a genome-wide assessment of the methylation present but do not give information on specific CpG dinucleotides and are biased toward CG rich regions of the genome.

PROGRESS TOWARD CHARACTERIZING THE METHYLOME OF LYMPHOMA AND LEUKEMIA

Genome-Wide Microarrays

Genome-wide methylation studies in lymphomas have progressed rapidly in light of the development of genome-wide technologies. Some of the earlier microarray platforms allowed for the investigation of DNA methylation present within select CpG islands [52,53]. The coverage of these arrays was limited but facilitated the discovery of putative tumor suppressor genes that are methylated in lymphomas and leukemias. Together with the identification of aberrantly methylated genes, investigators were able to differentiate between tumor types and subtypes [54,55,56,57,58,59] and to identify methylation hotspots in the genome [56,60]. The lessons learned from these studies prompted the development of commercial arrays that cover CpG islands, promoters of genes and even the whole genome (Agilent, Nimblegen Affymetrix, Illumina). These comprehensive arrays have not only allowed investigators to continue to distinguish between disease types, subtypes and outcomes [61,62,63,64,65,66,67,68,69] but have also allowed for the identification of enriched gene pathways and functional groups including PRC2 target genes [65,70,71] which has led to a better understanding of the biology of lymphomas and leukemias. These discoveries have profound implications with regards to prognosis, diagnosis and possibly even treatment of lymphomas and leukemias and highlight the importance of fully elucidating epigenomic changes in these disease types.

Ultradeep Bisulfite NGS of Amplicons in B-Cell Lymphomas

In 2007 our group developed a novel approach that allowed the simultaneous detection of methylation present in the CpG rich region of 25 genes in over 40 cases of primary cells [28]. To accomplish this goal, primer sets with disease specific tags on the 5' end were developed so that after NGS the groups could be identified (Figure 4A). Five groups were included, normal peripheral blood mononuclear cells (PBMCs), acute lymphoblastic leukemia (ALL), chronic lymphocytic leukemia (CLL), follicular lymphoma (FL) and mantle cell lymphoma (MCL). Ten individual samples within the same group were pooled together and then bisulfite treated. Gene specific primers containing disease specific tags were used to generate 25 amplicons for each group. Using clustering algorithms, differential methylation between diseases was shown. While this was not a genome-wide study it allowed for deep sequencing of amplicons representing genes whose aberrant methylation was previously shown to be methylated in lymphomas or other cancers [55,56,72,73]. Each amplicon was sequenced 100-1000 times providing improved statistical power over that of standard bisulfite sequencing. Twenty of the 25 amplicons studied showed an increase in methylation in various diseases as compared to PBMCs. Quantitative differences in methylation between diseases were also identified and there was a significant increase in the methylation of the 25 amplicons in ALL and FL samples compared to CLL and MCL. Unlike methods which provide regional methylation information, this method identified the precise location of the methylation present within the promoters of a subset of genes. A spreading of methylation was observed from the periphery towards the center of select CpG islands in ALL and FL samples. In addition to generating methylation profiles, deep bisulfite sequencing of amplicons allowed for the determination of an association between a SNP and methylation and also provided mechanistic insight with regards to the spreading of methylation.

RRBS/MIRA Plus Bisulfite Sequencing In A Follicular Lymphoma Cell Line

In addition to the published work described above, more recently we developed an approach integrating the concept of RRBS with MIRA plus bisulfite conversion to produce a library that was sequenced using the 454 GS FLX sequencer (Figure 4B). The inclusion of the enrichment step before bisulfite conversion allows for the targeting of methylated regions of the genome but is biased toward GC rich DNA sequences. Briefly, genomic DNA was fragmented using Csp6I and then ligated to adaptors, methylated adaptor-ligated fragments were enriched using a methyl-binding protein and the enriched fragments were subjected to bisulfite modification. A sequencing library was constructed and the follicular lymphoma cell line, RL, was sequenced. The genome-wide methylation patterns were compared with gene expression and histone modification profiles from the same cell line. More than 11,000 methylated regions of interest (MRI) were identified covering 4033 CpG islands. Methylation was detected in members of the HOX clustered genes, the protocadherin cluster, SOX genes and Frizzled protein genes and there was a significant overlap in methylation and PRC2 target genes. Genome-wide H3K4Me3, H3K27me3 and SUZ12 binding profiles were generated to compare to the MRIs identified by NGS. As expected, there was no significant overlap of methylation with H3K4 trimethylation. There was also no significant overlap of methylation and H3K27 trimethylation or SUZ12 binding *in vivo*. Furthermore, a SNP in the HLA-A gene cluster was associated with methylation status of 2 neighboring CpG sites. The longer bisulfite sequences obtained by 454 sequencing made sequence alignment accurate and reliable and allowed us to confirm a large number of methylated genes in RL cells and to identify distinguishing methylation patterns in the promoters of these genes.

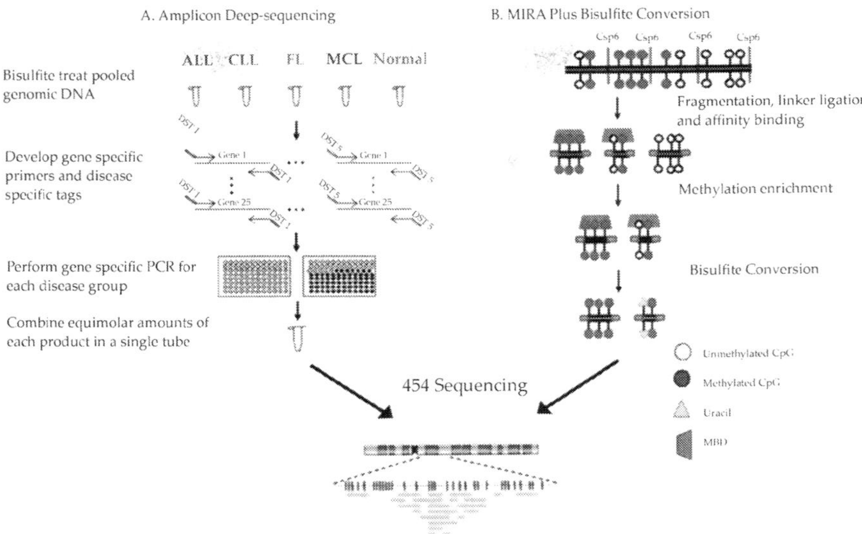

Figure 4. NGS approaches used to study lymphoid malignancies. **A.** Amplicon deep-sequencing. ALL: acute lymphoblastic leukemia; CLL: chronic lymphocytic leukemia; FL: follicular lymphoma; MCL: mantle cell lymphoma; DST: disease specific tag. **B.**MIRA plus bisulfite conversion.

We are at the forefront of exploiting NGS technologies to characterize the methylomes of lymphoid malignancies. The lessons learned and the information amassed from methylation microarrays is astounding but progress in this endeavor will increase exponentially now that we have the ability to produce true genome-scale data sets (Figure 5). The NGS technologies afford the opportunity to characterize the methylome much more deeply which is paramount to fully elucidating the mechanisms associated with, and the impacts of, aberrant methylation in lymphoid malignancies.

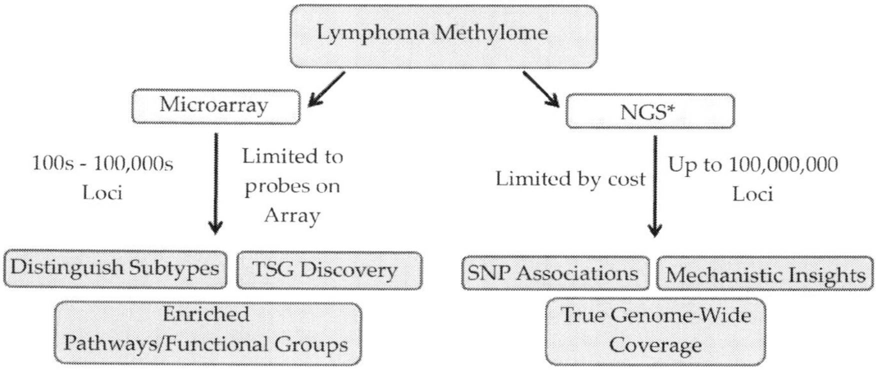

Figure 5. Microarray *versus* NGS in lymphoma methylation studies. NGS: Next-generation sequencing; TSG: tumor suppressor gene; SNP: single nucleotide polymorphism. * NGS includes all outcomes of microarray in addition to the outcomes presented in the diagram.

BEYOND THE METHYLOME

While DNA methylation was the focus of this review, there are other important epigenetic alterations which are also receiving deserving attention. For example, in normal cells, histone acetyltransferases (HATs) add acetyl groups to residues on histone tails that leads to a relaxation of the chromatin structure, allowing gene transcription to occur. HAT activity is counteracted by histone deacetylases (HDACs), enzymes that remove the acetyl groups leading to compaction of nucleosomes and transcriptional downregulation. Inhibitors of HDACs (HDACi) are thought to maintain chromatin in an open structure and are therefore potential pharmacologic agents because many tumors overexpress HDACs and this results in suppression of gene transcription (reviewed in [74]). A number of genome-wide studies have identified combinatorial patterns of histone modifications that may correlate with gene structures and functions [75,76,77,78]. These studies have mainly addressed the processes of acetylation or methylation at lysine (K) or arginine (R) residues in histone protein tails. Generally, trimethylation of histone H3K4 is associated with active transcription, while that of H3K27 is most commonly associated with downregulation, although both marks can appear simultaneously on bivalent gene promoters that may be poised for rapid expression or downregulation. Certain epigenetic modifications of chromatin by protein arginine methyltransferases (PRMTs) are also crucial for normal cell growth and differentiation. The mechanisms controlling this epigenetic process are not as well understood compared to lysine methylation. The SWI/SNF-associated protein PRMT5 transcriptionally represses its target genes by methylating histone 3 at arginine 8 (H3R8) and histone 4 at arginine 3 (H4R3) [79]. PRMT5 protein levels are increased in CLL cell lines secondary to the altered expression of PRMT5-specific miRs 19a, 25, 32, 92, 92b, and 96 (discussed below) that results in an increase of global symmetric methylation of H3R8 and H4R3. An evaluation of both epigenetic marks at PRMT5 target genes such as cell cycle pocket proteins RB1 (p105), RBL1 (p107), and RBL2 (p130) showed that at their promoters H3R8 and H4R3 are hypermethylated, which then leads to transcriptional repression. Using the human beta-globin locus as a model, symmetric dimethylation of H4R3me2 by PRMT5 is required for subsequent DNA methylation [79]. H4R3me2 is a direct binding target for DNMT3A, which interacts through the ADD domain containing the PHD motif. Loss of the H4R3me2 mark leads to reduced DNMT3A binding, loss of DNA methylation and gene activation. Thus, it was proposed that DNMT3A functions as both a reader and a writer of repressive epigenetic marks at this locus, thereby directly linking histone and DNA methylation in gene silencing.

Small RNA molecules, referred to as microRNA or miR, act as translational repressors of many mRNAs and each may affect hundreds of genes [80]. MiRs are processed after transcription by Drosha and Dicer enzymes and ultimately associate with an RNA-induced silencing complex (RISC) that binds target mRNAs through partially complementary sequences that then reduce their translation or stability. Human miR genes are distributed throughout the genome in a non-random manner and are frequently located in chromosomal fragile sites and cancer-associated genomic regions where they may act as molecular switches (reviewed in [81,82]). As with other coding genes, miR genes have specific functions, and are also susceptible to epigenetic modifications such as DNA hyper- and hypomethylation [83,84]. Some miRs play a fairly direct role in chromatin modification by targeting post-transcriptional regulation of key enzymes. For instance, the miR-29 family targets the three main DNA methyltransferases of humans either directly or indirectly in lung cancer and acute myeloid leukemias [20,21]. These miRs directly target the 3'UTR of DNMT3A and DNMT3B, but affect DNMT1 indirectly by targeting SP1 which functions as a transactivator of this enzyme gene. These enzymes are clearly important for DNA methylation in all cancers thus far studied.

ON THE HORIZON

As discussed, the epigenome is comprised of multiple modifications to the genome. Certainly, DNA methylation, histone modifications, and miRs are key players, but the global transcriptome is affected by many other protein-protein and protein-nucleic acid interactions that in concert can effect changes through overall chromatin organization, chromosomal territory localizations, and gene-gene interactions through loops of inter- and intra-chromosomal DNA. The chromatin state is an epigenomic mechanism for wide-spread nuclear activities. For instance, the patterns of euchromatin and heterochromatin differ between lymphoma subtypes, as evidenced by electron microscopy. In general, the transcriptionally active euchromatic regions are found near the center of the nucleus, while less active heterochromatic regions generally localize near the nuclear membrane. Even beyond the visual differences, with the aid of a powerful methods such as Hi-C and formaldehyde-assisted isolation of regulatory elements (FAIRE) [85] coupled with genomic sequencing (FAIRE-seq) [86], it can been shown that the actual genomic regions and associated genes found in either the euchromatic or heterochromatic regions of the nucleus also differ by cell type. Job Dekker's laboratory has developed multiple methods to probe genomic interactions. The most recent, Hi-C, is a method for defining the three-dimensional architecture of entire genomes through proximity-based ligation followed by massively parallel NGS. Through construction of spatial proximity maps of the human genome at a resolution of 1 megabase, they confirmed the presence of chromosome territories and the spatial proximity of small, gene-rich chromosomes, as well as an additional level of genome organization characterized by the spatial segregation of euchromatin and heterochromatin to form two genome-wide compartments [87]. Thus, Hi-C and NGS can be used to map the dynamic conformations of entire genomes. A complementary method, FAIRE-seq, can be used to investigate the genomic regions associated with the open chromatin regions of cells. Modifications to this method should also allow investigation of the protein-associated DNA found mainly in heterochromatic regions. It is inevitable that further refinements in NGS will facilitate studies at multiple levels of the epigenome and chromatin organization that can provide global information affecting the cellular transcriptome, proteome, and biological activity that defines clinical behaviors in all types of lymphoma, as well as other cancers and normal cells.

ACKNOWLEDGMENTS

Funding from National Institutes of Health grants CA123018, CA132706 and DA025779 (H. Shi), CA100055 and CA097880 (C. Caldwell), CA132784 (K. Taylor) and the CRC Missouri Chair in Cancer Research supported much of the data collection on which this review is based. We kindly thank Sam I. Hooshmand for assistance in the preparation of the figures included in this review. We also thank the reviewers for the thoughtful critiques that improved this manuscript.

REFERENCES AND NOTES

1. Egger, G.; Liang, G.; Aparicio, A.; Jones, P.A. Epigenetics in Human Disease and Prospects for Epigenetic Therapy. *Nature* **2004**, *429*, 457–463.
2. Feinberg, A.P.; Ohlsson, R.; Henikoff, S. The Epigenetic Progenitor Origin of Human Cancer. *Nat. Rev. Genet.* **2006**, *7*, 21–33.
3. Laird, P.W. The Power and the Promise of DNA Methylation Markers. *Nat. Rev. Cancer* **2003**, *3*, 253–266.
4. Bird, A. DNA Methylation Patterns and Epigenetic Memory. *Genes Dev.* **2002**, *16*, 6–21.
5. Robertson, K.D.; Jones, P.A. DNA Methylation: Past, Present and Future Directions. *Carcinogenesis* **2000**, *21*, 461–467.
6. Craig, J.M.; Bickmore, W.A. The Distribution of CpG Islands in Mammalian Chromosomes. *Nat. Genet.* **1994**, *7*, 376–382.
7. Goto, T.; Monk, M. Regulation of X-Chromosome Inactivation in Development in Mice and Humans. *Microbiol. Mol. Biol. Rev.* **1998**, *62*, 362–378.
8. Li, E.; Beard, C.; Jaenisch, R. Role for DNA Methylation in Genomic

Imprinting. *Nature* **1993**, *366*, 362–365.

9. Razin, A.; Cedar, H. DNA Methylation and Genomic Imprinting. *Cell* **1994**, *77*, 473–476.
10. Esteller, M. Cancer Epigenomics: DNA Methylomes and Histone-Modification Maps. *Nat. Rev. Genet.* **2007**, *8*, 286–298.
11. Gal-Yam, E.N.; Saito, Y.; Egger, G.; Jones, P.A. Cancer Epigenetics: Modifications, Screening, and Therapy. *Annu. Rev. Med.* **2008**, *59*, 267–280.
12. Irizarry, R.A.; Ladd-Acosta, C.; Wen, B.; Wu, Z.; Montano, C.; Onyango, P.; Cui, H.; Gabo, K.; Rongione, M.; Webster, M.; et al. The Human Colon Cancer Methylome Shows Similar Hypo- and Hypermethylation at Conserved Tissue-Specific CpG Island Shores . *Nat. Genet.* **2009**, *41*, 178–186.
13. Bracken, A.P.; Dietrich, N.; Pasini, D.; Hansen, K.H.; Helin, K. Genome-Wide Mapping of Polycomb Target Genes Unravels Their Roles in Cell Fate Transitions. *Genes Dev.* **2006**, *20*, 1123–1136.
14. Kirmizis, A.; Bartley, S.M.; Kuzmichev, A.; Margueron, R.; Reinberg, D.; Green, R.; Farnham, P.J. Silencing of Human Polycomb Target Genes Is Associated With Methylation of Histone H3 Lys 27. *Genes Dev.* **2004**, *18*, 1592–1605.
15. Lee, M.G.; Villa, R.; Trojer, P.; Norman, J.; Yan, K.P.; Reinberg, D.; Di, C.L.; Shiekhattar, R. Demethylation of H3K27 Regulates Polycomb Recruitment and H2A Ubiquitination. *Science* **2007**, *318*, 447–450.
16. Raaphorst, F.M. Deregulated Expression of Polycomb-Group Oncogenes in Human Malignant Lymphomas and Epithelial Tumors . *Hum. Mol. Genet.* **2005**, *14*, R93–R100.
17. Schlesinger, Y.; Straussman, R.; Keshet, I.; Farkash, S.; Hecht, M.; Zimmerman, J.; Eden, E.; Yakhini, Z.; Ben-Shushan, E.; Reubinoff, B.E.; Bergman, Y.; Simon, I.; Cedar, H. Polycomb-Mediated Methylation on Lys27 of Histone H3 Pre-Marks Genes for *De Novo* Methylation in Cancer. *Nat. Genet.* **2007**, *39*, 232–236.
18. Vire, E.; Brenner, C.; Deplus, R.; Blanchon, L.; Fraga, M.; Didelot, C.; Morey, L.; Van, E.A.; Bernard, D.; Vanderwinden, J.M.; et al. The Polycomb Group Protein EZH2 Directly Controls DNA Methylation . *Nature* **2005**, *439*, 871–874.

19. Ng, E.K.; Tsang, W.P.; Ng, S.S.; Jin, H.C.; Yu, J.; Li, J.J.; Rocken, C.; Ebert, M.P.; Kwok, T.T.; Sung, J.J. MicroRNA-143 Targets DNA Methyltransferases 3A in Colorectal Cancer . *Br. J. Cancer* **2009**, *101*, 699–706.

20. Fabbri, M.; Garzon, R.; Cimmino, A.; Liu, Z.; Zanesi, N.; Callegari, E.; Liu, S.; Alder, H.; Costinean, S.; Fernandez-Cymering, C.; et al. MicroRNA-29 Family Reverts Aberrant Methylation in Lung Cancer by Targeting DNA Methyltransferases 3A and 3B . *Proc. Natl. Acad. Sci. U. S. A.* **2007**, *104*, 15805–15810.

21. Garzon, R.; Liu, S.; Fabbri, M.; Liu, Z.; Heaphy, C.E.; Callegari, E.; Schwind, S.; Pang, J.; Yu, J.; Muthusamy, N.; et al. MicroRNA-29b Induces Global DNA Hypomethylation and Tumor Suppressor Gene Reexpression in Acute Myeloid Leukemia by Targeting Directly DNMT3A and 3B and Indirectly DNMT1 . *Blood* **2009**, *113*, 6411–6418.

22. Laird, P.W. Principles and Challenges of Genome-Wide DNA Methylation Analysis. *Nat. Rev. Genet.* **2010**, *11*, 191–203.

23. Lister, R.; Pelizzola, M.; Dowen, R.H.; Hawkins, R.D.; Hon, G.; Tonti-Filippini, J.; Nery, J.R.; Lee, L.; Ye, Z.; Ngo, Q.M.; et al. Human DNA Methylomes at Base Resolution Show Widespread Epigenomic Differences . *Nature* **2009**, *462*, 315–322.

24. Metzker, M.L. Sequencing Technologies - the Next Generation. *Nat. Rev. Genet.* **2010**, *11*, 31–46.

25. Mardis, E.R. The Impact of Next-Generation Sequencing Technology on Genetics. *Trends Genet.* **2008**, *24*, 133–141.

26. Korshunova, Y.; Maloney, R.K.; Lakey, N.; Citek, R.W.; Bacher, B.; Budiman, A.; Ordway, J.M.; McCombie, W.R.; Leon, J.; Jeddeloh, J.A.; McPherson, J.D. Massively Parallel Bisulphite Pyrosequencing Reveals the Molecular Complexity of Breast Cancer-Associated Cytosine-Methylation Patterns Obtained From Tissue and Serum DNA. *Genome Res.* **2008**, *18*, 19–29.

27. Ordway, J.M.; Budiman, M.A.; Korshunova, Y.; Maloney, R.K.; Bedell, J.A.; Citek, R.W.; Bacher, B.; Peterson, S.; Rohlfing, T.; Hall, J.; Brown, R.; Lakey, N.; Doerge, R.W.; Martienssen, R.A.; Leon, J.; McPherson, J.D.; Jeddeloh, J.A. Identification of Novel High-Frequency DNA Methylation Changes in Breast Cancer .

PLoS. One **2007**, *2*, e1314.

28. Taylor, K.H.; Kramer, R.S.; Davis, J.W.; Guo, J.; Duff, D.J.; Xu, D.; Caldwell, C.W.; Shi, H. Ultradeep Bisulfite Sequencing Analysis of DNA Methylation Patterns in Multiple Gene Promoters by 454 Sequencing. *Cancer Res.* **2007**, *67*, 8511–8518.
29. Xie, H.; Wang, M.; Bonaldo, M.F.; Rajaram, V.; Stellpflug, W.; Smith, C.; Arndt, K.; Goldman, S.; Tomita, T.; Soares, M.B. Epigenomic Analysis of Alu Repeats in Human Ependymomas. *Proc. Natl. Acad. Sci. U. S. A.* **2010**, *107*, 6952–6957.
30. Deng, J.; Shoemaker, R.; Xie, B.; Gore, A.; LeProust, E.M.; ntosiewicz-Bourget, J.; Egli, D.; Maherali, N.; Park, I.H.; Yu, J.; Daley, G.Q.; Eggan, K.; Hochedlinger, K.; Thomson, J.; Wang, W.; Gao, Y.; Zhang, K. Targeted Bisulfite Sequencing Reveals Changes in DNA Methylation Associated With Nuclear Reprogramming. *Nat. Biotechnol.* **2009**, *27*, 353–360.
31. Gu, H.; Bock, C.; Mikkelsen, T.S.; Jager, N.; Smith, Z.D.; Tomazou, E.; Gnirke, A.; Lander, E.S.; Meissner, A. Genome-Scale DNA Methylation Mapping of Clinical Samples at Single-Nucleotide Resolution. *Nat. Methods* **2010**, *7*, 133–136.
32. Hodges, E.; Smith, A.D.; Kendall, J.; Xuan, Z.; Ravi, K.; Rooks, M.; Zhang, M.Q.; Ye, K.; Bhattacharjee, A.; Brizuela, L.; et al. High Definition Profiling of Mammalian DNA Methylation by Array Capture and Single Molecule Bisulfite Sequencing . *Genome Res.* **2009**, *19*, 1593–1605.
33. Cokus, S.J.; Feng, S.; Zhang, X.; Chen, Z.; Merriman, B.; Haudenschild, C.D.; Pradhan, S.; Nelson, S.F.; Pellegrini, M.; Jacobsen, S.E. Shotgun Bisulphite Sequencing of the Arabidopsis Genome Reveals DNA Methylation Patterning. *Nature* **2008**, *452*, 215–219.
34. Feng, S.; Cokus, S.J.; Zhang, X.; Chen, P.Y.; Bostick, M.; Goll, M.G.; Hetzel, J.; Jain, J.; Strauss, S.H.; Halpern, M.E.; Ukomadu, C.; Sadler, K.C.; Pradhan, S.; Pellegrini, M.; Jacobsen, S.E. Conservation and Divergence of Methylation Patterning in Plants and Animals. *Proc. Natl. Acad. Sci. U. S. A.* **2010**, *107*, 8689–8694.
35. Lister, R.; O'Malley, R.C.; Tonti-Filippini, J.; Gregory, B.D.; Berry, C.C.; Millar, A.H.; Ecker, J.R. Highly Integrated Single-Base

Resolution Maps of the Epigenome in Arabidopsis. *Cell* **2008**, *133*, 523–536.

36. Ball, M.P.; Li, J.B.; Gao, Y.; Lee, J.H.; LeProust, E.M.; Park, I.H.; Xie, B.; Daley, G.Q.; Church, G.M. Targeted and Genome-Scale Strategies Reveal Gene-Body Methylation Signatures in Human Cells. *Nat. Biotechnol.* **2009**, *27*, 361–368.
37. Li, J.; Gao, F.; Li, N.; Li, S.; Yin, G.; Tian, G.; Jia, S.; Wang, K.; Zhang, X.; Yang, H.; Nielsen, A.L.; Bolund, L. An Improved Method for Genome Wide DNA Methylation Profiling Correlated to Transcription and Genomic Instability in Two Breast Cancer Cell Lines. *BMC Genomics* **2009**, *10*, 223.
38. Shendure, J.; Porreca, G.J.; Reppas, N.B.; Lin, X.; McCutcheon, J.P.; Rosenbaum, A.M.; Wang, M.D.; Zhang, K.; Mitra, R.D.; Church, G.M. Accurate Multiplex Polony Sequencing of an Evolved Bacterial Genome. *Science* **2005**, *309*, 1728–1732.
39. Bormann Chung, C.A.; Boyd, V.L.; McKernan, K.J.; Fu, Y.; Monighetti, C.; Peckham, H.E.; Barker, M. Whole Methylome Analysis by Ultra-Deep Sequencing Using Two-Base Encoding . *PLoS One* **2010**, *5*, e9320.
40. Drmanac, R.; Sparks, A.B.; Callow, M.J.; Halpern, A.L.; Burns, N.L.; Kermani, B.G.; Carnevali, P.; Nazarenko, I.; Nilsen, G.B.; Yeung, G.; et al. Human Genome Sequencing Using Unchained Base Reads on Self-Assembling DNA Nanoarrays . *Science* **2010**, *327*, 78–81.
41. Laurent, L.; Wong, E.; Li, G.; Huynh, T.; Tsirigos, A.; Ong, C.T.; Low, H.M.; Kin Sung, K.W.; Rigoutsos, I.; Loring, J.; Wei, C.L. Dynamic Changes in the Human Methylome During Differentiation. *Genome Res.* **2010**, *20*, 320–331.
42. Meissner, A.; Gnirke, A.; Bell, G.W.; Ramsahoye, B.; Lander, E.S.; Jaenisch, R. Reduced Representation Bisulfite Sequencing for Comparative High-Resolution DNA Methylation Analysis. *Nucleic Acids Res.* **2005**, *33*, 5868–5877.
43. Smith, Z.D.; Gu, H.; Bock, C.; Gnirke, A.; Meissner, A. High-Throughput Bisulfite Sequencing in Mammalian Genomes. *Methods* **2009**, *48*, 226–232.
44. Zeschnigk, M.; Martin, M.; Betzl, G.; Kalbe, A.; Sirsch, C.; Buiting,

K.; Gross, S.; Fritzilas, E.; Frey, B.; Rahmann, S.; Horsthemke, B. Massive Parallel Bisulfite Sequencing of CG-Rich DNA Fragments Reveals That Methylation of Many X-Chromosomal CpG Islands in Female Blood DNA Is Incomplete. *Hum. Mol. Genet.* **2009**, *18*, 1439–1448.

45. Birney, E.; Stamatoyannopoulos, J.A.; Dutta, A.; Guigo, R.; Gingeras, T.R.; Margulies, E.H.; Weng, Z.; Snyder, M.; Dermitzakis, E.T.; Thurman, R.E.; et al. Identification and Analysis of Functional Elements in 1% of the Human Genome by the ENCODE Pilot Project . *Nature* **2007**, *447*, 799–816.
46. Oda, M.; Glass, J.L.; Thompson, R.F.; Mo, Y.; Olivier, E.N.; Figueroa, M.E.; Selzer, R.R.; Richmond, T.A.; Zhang, X.; Dannenberg, L.; et al. High-Resolution Genome-Wide Cytosine Methylation Profiling With Simultaneous Copy Number Analysis and Optimization for Limited Cell Numbers . *Nucleic Acids Res.* **2009**, *37*, 3829–3839.
47. Wang, X.; Elling, A.A.; Li, X.; Li, N.; Peng, Z.; He, G.; Sun, H.; Qi, Y.; Liu, X.S.; Deng, X.W. Genome-Wide and Organ-Specific Landscapes of Epigenetic Modifications and Their Relationships to MRNA and Small RNA Transcriptomes in Maize. *Plant Cell* **2009**, *21*, 1053–1069.
48. Rauch, T.; Pfeifer, G.P. Methylated-CpG Island Recovery Assay: a New Technique for the Rapid Detection of Methylated-CpG Islands in Cancer. *Lab. Invest.* **2005**, *85*, 1172–1180.
49. Weber, M.; Davies, J.J.; Wittig, D.; Oakeley, E.J.; Haase, M.; Lam, W.L.; Schubeler, D. Chromosome-Wide and Promoter-Specific Analyses Identify Sites of Differential DNA Methylation in Normal and Transformed Human Cells. *Nat. Genet.* **2005**, *37*, 853–862.
50. Butcher, L.M.; Beck, S. AutoMeDIP-Seq: A High-Throughput, Whole Genome, DNA Methylation Assay . *Methods*. 2010. Available online: http://www.ncbi.nlm.nih.gov/pubmed/20385236/.
51. Serre, D.; Lee, B.H.; Ting, A.H. MBD-Isolated Genome Sequencing Provides a High-Throughput and Comprehensive Survey of DNA Methylation in the Human Genome. *Nucleic Acids Res.* **2010**, *38*, 391–399.
52. Heisler, L.E.; Torti, D.; Boutros, P.C.; Watson, J.; Chan, C.; Winegarden, N.; Takahashi, M.; Yau, P.; Huang, T.H.; Farnham,

P.J.; Jurisica, I.; Woodgett, J.R.; Bremner, R.; Penn, L.Z.; Der, S.D. CpG Island Microarray Probe Sequences Derived From a Physical Library Are Representative of CpG Islands Annotated on the Human Genome. *Nucleic Acids Res.* **2005**, *33*, 2952–2961.

53. Yan, P.S.; Chen, C.M.; Shi, H.; Rahmatpanah, F.; Wei, S.H.; Caldwell, C.W.; Huang, T.H. Dissecting Complex Epigenetic Alterations in Breast Cancer Using CpG Island Microarrays. *Cancer Res.* **2001**, *61*, 8375–8380.
54. Guo, J.; Burger, M.; Nimmrich, I.; Maier, S.; Becker, E.; Genc, B.; Duff, D.; Rahmatpanah, F.; Chitma-Matsiga, R.; Shi, H.; Berlin, K.; Huang, T.H.; Caldwell, C.W. Differential DNA Methylation of Gene Promoters in Small B-Cell Lymphomas. *Am. J. Clin. Pathol.* **2005**, *124*, 430–439.
55. Shi, H.; Guo, J.; Duff, D.J.; Rahmatpanah, F.; Chitima-Matsiga, R.; Al-Kuhlani, M.; Taylor, K.H.; Sjahputera, O.; Andreski, M.; Wooldridge, J.E.; Caldwell, C.W. Discovery of Novel Epigenetic Markers in Non-Hodgkin's Lymphoma. *Carcinogenesis* **2007**, *28*, 60–70.
56. Taylor, K.H.; Pena-Hernandez, K.E.; Davis, J.W.; Arthur, G.L.; Duff, D.J.; Shi, H.; Rahmatpanah, F.B.; Sjahputera, O.; Caldwell, C.W. Large-Scale CpG Methylation Analysis Identifies Novel Candidate Genes and Reveals Methylation Hotspots in Acute Lymphoblastic Leukemia. *Cancer Res.* **2007**, *67*, 2617–2625.
57. van, D.R.; Zoutman, W.H.; Dijkman, R.; de Menezes, R.X.; Commandeur, S.; Mulder, A.A.; van, d.; Vermeer, M.H.; Willemze, R.; Yan, P.S.; Huang, T.H.; Tensen, C.P. Epigenetic Profiling of Cutaneous T-Cell Lymphoma: Promoter Hypermethylation of Multiple Tumor Suppressor Genes Including BCL7a, PTPRG, and P73 . *J. Clin. Oncol.* **2005**, *23*, 3886–3896.
58. Scholz, C.; Nimmrich, I.; Burger, M.; Becker, E.; Dorken, B.; Ludwig, W.D.; Maier, S. Distinction of Acute Lymphoblastic Leukemia From Acute Myeloid Leukemia Through Microarray-Based DNA Methylation Analysis. *Ann. Hematol.* **2005**, *84*, 236–244.
59. Martin-Subero, J.I.; Ammerpohl, O.; Bibikova, M.; Wickham-Garcia, E.; Agirre, X.; Alvarez, S.; Bruggemann, M.; Bug, S.;

Calasanz, M.J.; Deckert, M.; et al. A Comprehensive Microarray-Based DNA Methylation Study of 367 Hematological Neoplasms . *PLoS. One.* **2009**, *4*, e6986.

60. Taylor, K.H.; Rahmatpanah, F.; Davis, J.W.; Caldwell, C.W. Chromosomal Localization of DNA Methylation in Small B-Cell Lymphoma. *Leukemia* **2007**, *22*, 638–641.
61. Davidsson, J.; Lilljebjorn, H.; Andersson, A.; Veerla, S.; Heldrup, J.; Behrendtz, M.; Fioretos, T.; Johansson, B. The DNA Methylome of Pediatric Acute Lymphoblastic Leukemia. *Hum. Mol. Genet.* **2009**, *18*, 4054–4065.
62. Figueroa, M.E.; Reimers, M.; Thompson, R.F.; Ye, K.; Li, Y.; Selzer, R.R.; Fridriksson, J.; Paietta, E.; Wiernik, P.; Green, R.D.; Greally, J.M.; Melnick, A. An Integrative Genomic and Epigenomic Approach for the Study of Transcriptional Regulation . *PLoS One* **2008**, *3*, e1882.
63. Figueroa, M.E.; Wouters, B.J.; Skrabanek, L.; Glass, J.; Li, Y.; Erpelinck-Verschueren, C.A.; Langerak, A.W.; Lowenberg, B.; Fazzari, M.; Greally, J.M.; Valk, P.J.; Melnick, A.; Delwel, R. Genome-Wide Epigenetic Analysis Delineates a Biologically Distinct Immature Acute Leukemia With Myeloid/T-Lymphoid Features. *Blood* **2009**, *113*, 2795–2804.
64. Kanduri, M.; Cahill, N.; Goransson, H.; Enstrom, C.; Ryan, F.; Isaksson, A.; Rosenquist, R. Differential Genome-Wide Array-Based Methylation Profiles in Prognostic Subsets of Chronic Lymphocytic Leukemia. *Blood* **2010**, *115*, 296–305.
65. Martin-Subero, J.I.; Kreuz, M.; Bibikova, M.; Bentink, S.; Ammerpohl, O.; Wickham-Garcia, E.; Rosolowski, M.; Richter, J.; Lopez-Serra, L.; Ballestar, E.; et al. New Insights into the Biology and Origin of Mature Aggressive B-Cell Lymphomas by Combined Epigenomic, Genomic, and Transcriptional Profiling . *Blood* **2009**, *113*, 2488–2497.
66. Deneberg, S.; Grovdal, M.; Karimi, M.; Jansson, M.; Nahi, H.; Corbacioglu, A.; Gaidzik, V.; Dohner, K.; Paul, C.; Ekstrom, T.J.; Hellstrom-Lindberg, E.; Lehmann, S. Gene-Specific and Global Methylation Patterns Predict Outcome in Patients With Acute Myeloid Leukemia. *Leukemia* **2010**, *24*, 932–941.

67. Dunwell, T.; Hesson, L.; Rauch, T.A.; Wang, L.; Clark, R.E.; Dallol, A.; Gentle, D.; Catchpoole, D.; Maher, E.R.; Pfeifer, G.P.; Latif, F. A Genome-Wide Screen Identifies Frequently Methylated Genes in Haematological and Epithelial Cancers. *Mol. Cancer* **2010**, *9*, 44.

68. Schafer, E.; Irizarry, R.; Negi, S.; McIntyre, E.; Small, D.; Figueroa, M.E.; Melnick, A.; Brown, P. Promoter Hypermethylation in MLL-r Infant Acute Lymphoblastic Leukemia: Biology and Therapeutic Targeting. *Blood* **2010**, *115*, 4798–4809.

69. Shaknovich, R.; Figueroa, M.E.; Melnick, A. HELP (HpaII Tiny Fragment Enrichment by Ligation-Mediated PCR) Assay for DNA Methylation Profiling of Primary Normal and Malignant B Lymphocytes. *Methods Mol. Biol.* **2010**, *632*, 191–201.

70. Bennett, L.B.; Schnabel, J.L.; Kelchen, J.M.; Taylor, K.H.; Guo, J.; Arthur, G.L.; Papageorgio, C.N.; Shi, H.; Caldwell, C.W. DNA Hypermethylation Accompanied by Transcriptional Repression in Follicular Lymphoma. *Genes Chromosomes Cancer* **2009**, *48*, 828–841.

71. Kuang, S.Q.; Bai, H.; Fang, Z.H.; Lopez, G.; Yang, H.; Tong, W.; Wang, Z.Z.; Garcia-Manero, G. Aberrant DNA Methylation and Epigenetic Inactivation of Eph Receptor Tyrosine Kinases and Ephrin Ligands in Acute Lymphoblastic Leukemia. *Blood* **2010**, *115*, 2412–2419.

72. Esteller, M. Profiling Aberrant DNA Methylation in Hematologic Neoplasms: a View From the Tip of the Iceberg. *Clin. Immunol.* **2003**, *109*, 80–88.

73. Rahmatpanah, F.B.; Carstens, S.; Guo, J.; Sjahputera, O.; Taylor, K.H.; Duff, D.; Shi, H.; Davis, J.W.; Hooshmand, S.I.; Chitma-Matsiga, R.; Caldwell, C.W. Differential DNA Methylation Patterns of Small B-Cell Lymphoma Subclasses With Different Clinical Behavior. *Leukemia* **2006**, *20*, 1855–1862.

74. Lane, A.A.; Chabner, B.A. Histone Deacetylase Inhibitors in Cancer Therapy. *J. Clin. Oncol.* **2009**, *27*, 5459–5468.

75. Barski, A.; Cuddapah, S.; Cui, K.; Roh, T.Y.; Schones, D.E.; Wang, Z.; Wei, G.; Chepelev, I.; Zhao, K. High-Resolution Profiling of Histone Methylations in the Human Genome. *Cell* **2007**, *129*, 823–837.

76. Bernstein, B.E.; Kamal, M.; Lindblad-Toh, K.; Bekiranov, S.; Bailey, D.K.; Huebert, D.J.; McMahon, S.; Karlsson, E.K.; Kulbokas III, E.J.; Gingeras, T.R.; Schreiber, S.L.; Lander, E.S. Genomic Maps and Comparative Analysis of Histone Modifications in Human and Mouse . *Cell* **2005**, *120*, 169–181.
77. Bernstein, B.E.; Meissner, A.; Lander, E.S. The Mammalian Epigenome. *Cell* **2007**, *128*, 669–681.
78. Mikkelsen, T.S.; Ku, M.; Jaffe, D.B.; Issac, B.; Lieberman, E.; Giannoukos, G.; Alvarez, P.; Brockman, W.; Kim, T.K.; Koche, R.P.; et al. Genome-Wide Maps of Chromatin State in Pluripotent and Lineage-Committed Cells . *Nature* **2007**, *448*, 553–560.
79. Zhao, Q.; Rank, G.; Tan, Y.T.; Li, H.; Moritz, R.L.; Simpson, R.J.; Cerruti, L.; Curtis, D.J.; Patel, D.J.; Allis, C.D.; Cunningham, J.M.; Jane, S.M. PRMT5-Mediated Methylation of Histone H4R3 Recruits DNMT3A, Coupling Histone and DNA Methylation in Gene Silencing. *Nat. Struct. Mol. Biol.* **2009**, *16*, 304–311.
80. Ruan, K.; Fang, X.; Ouyang, G. MicroRNAs: Novel Regulators in the Hallmarks of Human Cancer. *Cancer Lett.* **2009**, *285*, 116–126.
81. Garzon, R.; Calin, G.A.; Croce, C.M. MicroRNAs in Cancer. *Annu. Rev. Med.* **2009**, *60*, 167–179.
82. Sotiropoulou, G.; Pampalakis, G.; Lianidou, E.; Mourelatos, Z. Emerging Roles of MicroRNAs As Molecular Switches in the Integrated Circuit of the Cancer Cell. *RNA* **2009**, *15*, 1443–1461.
83. Weber, B.; Stresemann, C.; Brueckner, B.; Lyko, F. Methylation of Human MicroRNA Genes in Normal and Neoplastic Cells. *Cell Cycle* **2007**, *6*, 1001–1005. [Google Scholar]
84. Guil, S.; Esteller, M. DNA Methylomes, Histone Codes and MiRNAs: Tying It All Together. *Int. J. Biochem. Cell Biol.* **2009**, *41*, 87–95.
85. Giresi, P.G.; Kim, J.; McDaniell, R.M.; Iyer, V.R.; Lieb, J.D. FAIRE (Formaldehyde-Assisted Isolation of Regulatory Elements) Isolates Active Regulatory Elements From Human Chromatin. *Genome Res.* **2007**, *17*, 877–885.
86. Gaulton, K.J.; Nammo, T.; Pasquali, L.; Simon, J.M.; Giresi, P.G.; Fogarty, M.P.; Panhuis, T.M.; Mieczkowski, P.; Secchi, A.; Bosco, D.; Berney, T.; Montanya, E.; Mohlke, K.L.; Lieb, J.D.; Ferrer, J.

A Map of Open Chromatin in Human Pancreatic Islets. *Nat. Genet.* **2010**, *42*, 255–259.

87. Lieberman-Aiden, E.; van Berkum, N.L.; Williams, L.; Imakaev, M.; Ragoczy, T.; Telling, A.; Amit, I.; Lajoie, B.R.; Sabo, P.J.; Dorschner, M.O.; et al. Comprehensive Mapping of Long-Range Interactions Reveals Folding Principles of the Human Genome . *Science* **2009**, *326*, 289–293.

CHAPTER 19

ANGSD: ANALYSIS OF NEXT GENERATION SEQUENCING DATA

Thorfinn Sand Korneliussen[1], Anders Albrechtsen[2] and Rasmus Nielsen[1,3]

[1]Centre for GeoGenetics, Natural History Museum of Denmark, Copenhagen, Denmark

[2]Bioinformatics Centre, Department of Biology, University of Copenhagen, Ole Maaloes Vej 5, DK-2200 Copenhagen, Denmark

[3]Department of Integrative Biology and Statistics, UC-Berkeley, 4098 VLSB, Berkeley, California 94720, USA

ABSTRACT

Background

High-throughput DNA sequencing technologies are generating vast amounts of data. Fast, flexible and memory efficient implementations are needed in order to facilitate analyses of thousands of samples simultaneously.

Results

We present a multithreaded program suite called ANGSD. This program can calculate various summary statistics, and perform association mapping and population genetic analyses utilizing the full information in next generation sequencing data by working directly on the raw sequencing data or by using genotype likelihoods.

Conclusions

The open source c/c++ program ANGSD is available at http://www.popgen.dk/angsd. The program is tested and validated on GNU/Linux systems. The program facilitates multiple input formats including BAM and imputed beagle genotype probability files. The program allow the user to choose between combinations of existing methods and can perform analysis that is not implemented elsewhere.

Keywords

Next-generation sequencing, Bioinformatics, Population genetics, Association studies

BACKGROUND

Next generation sequencing (NGS) platforms can generate large amounts of sequencing data, but often with high sequence error rates. For low to medium depth data fast and efficient implementation are needed to handle the data. Arguably, downstream analyses should be performed in a probabilistic context by working with the raw data in form of genotype likelihoods (GL) [1]. ANGSD is a novel and efficient program that allows for multiple error models used within the GL calculation. The remainder of this section describe the typical work flow used for analyzing data. The implementation section lists and describes existing (published) methods and new methods that are available in our tool. The majority of methods in ANGSD are not implemented in other software and in the results section we have therefore limited the comparisons with existing tools to basic analyses of SNP-discovery and genotype calling.

The first step in a bioinformatic pipeline for analyzing NGS data is usually to align the reads to a reference genome using a fast short read aligner [2]-[5]. State-of-the-art alignment programs will, in addition to inferring the genomic start position of the reads, provide additional information such as the mapping quality scores (mapQ), and possibly also indicate which parts of an alignment may be affected by indels. Information regarding sequencing quality is included in quality scores (qscores), typically provided by the sequencing technology, and often modified using downstream re-calibration [6]-[8]. Based on the aligned reads, and associated mapping and sequencing quality scores, a genotype likelihood (GL) is then calculated. The GL is (up to a scaling factor) the marginal probability of the sequencing data given a genotype in a particular individual, in a particular site. Most data analyses then proceed by calling SNPs and genotypes from the GLs, typically combining information from multiple individuals, often also combining the GL with prior information, such as the inferred distribution of allele frequencies. For many applications based on high-quality deeply sequenced data, this is a near-optimal strategy for analyzing the data. However, for low or medium coverage data, there is often a distinct statistical advantage in working on the raw data, or GLs, rather than called genotypes in downstream analyses [9]-[14]. Working directly on GLs facilitates the incorporation of statistical uncertainty regarding genotypes. The uncertainty regarding genotypes in low coverage data arises from several sources, including mapping and sequencing errors, and the random sampling of (haploid) reads from a diploid genotype.

The de facto standard format to store and distribute NGS data in, is the BAM format which allows for random access within the sequencing data. When analyzing many individuals simultaneously, due to memory constraints, it is often convenient to analyze regions or single sites independently instead of reading all the data into memory. This is achieved by reading parts of each BAM file, aligning and then passing the aligned sites for analysis. Here we present an open source mutithreaded C/C++ program called ANGSD with this capability. ANGSD provides easy user access to methods for population genetic analyses and association mapping utilizing the full information of the data and taking uncertainty regarding SNP calling and genotype calling into account, by working directly on user-provided, or de novo estimated, GLs.

Examples of existing general multisample NGS analysis programs are the single threaded SAMtools [15] (C) and the multithreaded GATK [8] (Java). There are many differences between the three programs, but the key advantage with ANGSD is that it 1) allows for multiple input data types relating directly to raw sequencing data (text mpileup, binary genotype likelihood files, VCF files), 2) allows the user to choose between multiple methods for intermediate analysis such as different ways to calculate GL and 3) includes implementations of a large set of downstream analyses not implemented in any other software.

IMPLEMENTATION

Input Formats

ANGSD can currently parse a variety of different input formats including binary BAM files and mpileup text files. Genotype likelihoods input are supported for simple genotype likelihood formats and it also supports genotype (posterior) probabilities in the BEAGLE [16] format. ANGSD can perform various analyses, but some of these can be limited by the chosen input format e.g., sequencing depth calculation can only be performed on the basis of raw sequencing files and not GLs. The dependency between the different analysis and input formats is depicted in Figure 1. Indexed BAM files facilitate random access and this feature is implemented in ANGSD. Random access is not supported for other file formats.

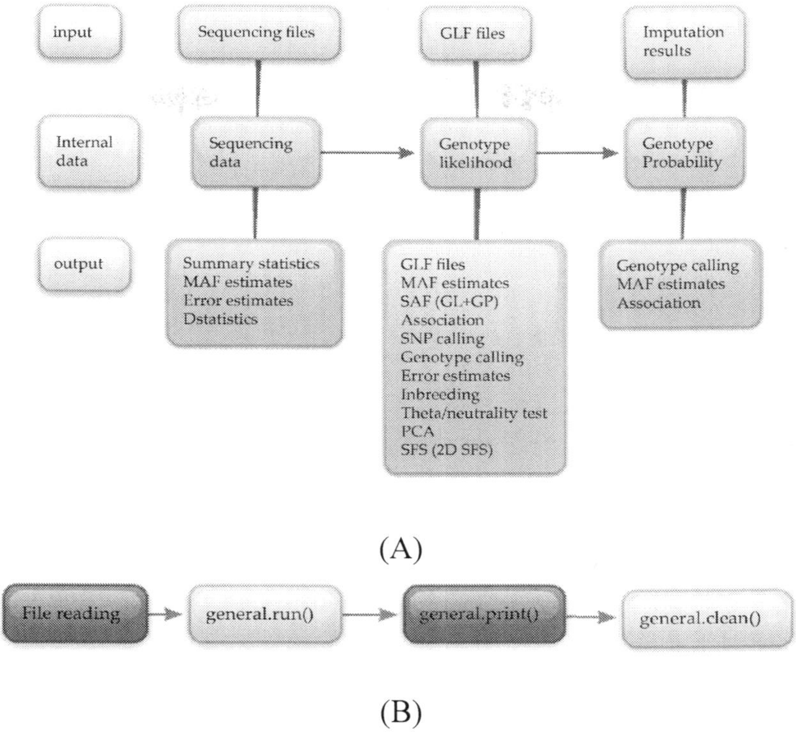

Figure 1: Data formats and call graph. A) Dependency of different data formats and analyses that can be performed in ANGSD. **B)** Simplified call graph. Red nodes indicate areas that are not threaded. With the exception of file readers, all analyses, printing and cleaning is done by objects derived from the abstract base class called general.

Methods

Some analyses can be performed on a single-site basis such as simple allele frequency estimation (MAF) and site-wise association testing. For analyses on genome wide data, the work-flow is divided into two steps: 1) ANGSD generates specific input data for the analysis. 2) A secondary associated program is used to perform the downstream analysis based on the ANGSD output [17]. For simple tests such as ABBA-BABA/D-statistic [18] the secondary program can be a simple Rscript, but for computational intensive methods it can be a multithreaded c/c++ program. A description of the main methods are found in Table 1. ANGSD allows a number of different types of input data, but not all analyses can be performed if the input data is not sequencing data. For example, only a few methods are applicable if the input data is genotype probabilities, e.g., likelihood ratio test for SNP calling can only be performed on GLs and not genotype probabilities (posteriors probabilities) (see Figure 1).

Table 1: Overview of analyses implemented in ANGSD

Analysis	Basis	Reference
Contamination estimates based on the X-chromosomes	BC	[19] [b]
Type specific error estimation estimated by simultaneously estimating allele frequencies and genotype likelihoods	GL	[10]
Type specific error estimation based on an outgroup and a high quality genome	BC	[20] [a b]
Genotype likelihoods (GL) (diploids)	BC/Seq	[6],[8],[10],[15]
Allele frequencies for a site	BC/GL/GP	[21] [b] [10]
SNP discovery (LRT) used for rejecting that the allele frequency is different from zero	GL	[10]
Genotype posteriors (GP) can be used for calling genotypes by specifying a cutoff	GL/SAF	[9],[10]
Sample allele frequencies (SAF) the probability of all read data given the sample allele frequency	GL/GP	[9] [b]
Population differentiation statistics $\boldsymbol{F_{st}}$	SAF	[14] [a c]
Population structure via principle components analysis **(PCA)**	GP	[14] [a c]
Admixture analysis (NGSadmix) NGS data	GL	[22] [a b]
Detection of ancient admixture **ABBA-BABA/d-statistics**	BC	[20] [b]
Estimation of **SFS (1D)**	SAF	[9] [a b]

Estimation of **SFS (2D)**	SAF	
Selection scans, Neutrality tests (e.g θ's and Tajima's D)	SAF	[12] [a b]
Estimation of individual and site-wise **Inbreeding** coefficients. Also MAF and GP estimation for inbreed individuals	GL	[13] [a b c]
Allele frequency based association for case/control data)	GL	[10]
Association score test in a generalized linear model framework for both quantitative and case/control data while allowing for additional covariates	GL-GP	[11] [b]

Table of the supported analyses in ANGSD. [a] indicates methods that require a secondary program in ANGSD package. [b]indicates methods for which ANGSD is the de facto implementation and [c] are user supplied extensions for ANGSD. The basis for each analysis is either the sequencing data (Seq), base counts (BC), genotype likelihood (GL), sample allele frequencies (SAF) or genotype probabilities (GP).

Genotype Likelihoods

For low and medium coverage NGS data, the recommended practice is to avoid basing downstream analysis on the raw counts of sequenced bases or called genotypes [1], but instead use a probabilistic approach by using GLs. Many of the methods within ANGSD are based on GLs (Table 1). ANGSD supports four different models for calculating GLs: 1) The recalibrating SOAPsnp model [6]. 2) The original GATK model [8] 3) SAMtools 1.16+ modified Maq model [23]. 4) The type specific error model [10]. The sequencing error rates in these GL models are either fixed, obtained from qscores, or estimated from the data. The four implemented GL models assume diploid samples.

Allele Frequency Estimators

The sample allele frequency in a site is the frequency of the allele among the individuals included in a specific sample. The population allele frequency is the (unknown) frequency of the allele in the entire population. Without genotype uncertainty, the sample allele frequency is known and the population allele frequency can be estimated from the sample allele frequency. However, in the presence of genotype uncertainty, the sample allele frequency is unknown, but can be estimated from the raw data or from the genotype likelihoods [9]. We have implemented several estimators of population (e.g., [10],[21]) and sample allele frequencies (e.g., [9]), that can be estimated based on GL's, base counts or genotype posteriors. By using the population allele frequency we have implemented a likelihood ratio test (LRT) of the site being variable which can be used as a SNP discovery criterion, and a Bayesian approach for calling genotypes.

Population Genetic Analysis Based On Sample Allele Frequencies

Several analyses are based on sample allele frequency likelihoods instead of single individual genotype likelihoods. A sample allele frequency likelihood is (up to a scaling factor) the probability of all read data for multiple individuals at a site, given the sample allele frequency. The methods in [9] use the sample allele frequency likelihood in several applications, including estimation of the site frequency spectrum (SFS), and estimation of Tajima's D and various other neutrality tests can be estimated taking genotype uncertainty into account [12]. These methods are included in the ANGSD package as separate programs that utilize ANGSD output. Various Bayesian estimation procedures are also implemented, including maximum a posteriori probability (MAP) estimates of the sample allele frequency [9]. The implementation in ANGSD allows for the use of externally estimated posterior probabilities (obtained for example using haplotype imputation based methods) for the calculation of posterior sample allele frequencies and other downstream analyses. Importantly, ANGSD also allows for the joint estimation of sample allele frequencies from two populations (2D-SFS):

Assuming two populations with n_1 and n_2 diploid individuals sampled from population 1 and 2, respectively. Then the 2D-SFS is the matrix γ :$(2n_1+1)\times(2n_2+1)$ of frequencies of derived sample allele counts in the two populations, i.e. γ_{ij} is the probability of observing i and j derived alleles population 1 and 2, respectively, in a randomly chosen site.

Let $p\left(X_s^d \mid D_d = i\right)$ denote the likelihood for the sequencing data, in population d for site s, given a total of I derived alleles in population d. This likelihood is calculated using the algorithm described in ref. [9]. We can then write the likelihood for a single site s for the 2D-SFS as:

$$L(X|\gamma) = \prod_{s=0}^{N} L(X_s \mid \gamma) = \prod_{s=0}^{N} \sum_{i=0}^{2n_1} \sum_{j=0}^{2n_2} \gamma_{ij} p\left(X_s^1 \mid D^1 = i\right) \times p\left(X_s^2 \mid D^2 = j\right) \quad (1)$$

In order to find the maximum likelihood we use an EM-algorithm. Assuming γ^{old} is our current parameters, a next iteration in the EM-algorithm is given by:

$$\gamma_{ij}^{new} = \sum_{s=0}^{N} p\left(X_s^1 \mid D^1 = i\right) p\left(X_s^2 \mid D^2 = j\right) / L\left(X_s \mid \gamma^{old}\right).$$

The algorithm then iterates updates of all ij simultaneously until the difference in successive likelihood values is below some tolerance.

Population Structure

Genomes for admixed individuals represent a mixture of alleles from different ancestral populations. Inferring individual admixture proportions along with a frequency estimate for the different ancestral populations is possible based on genotype likelihoods [22] based on output from ANGSD. Similarly the sample allele frequency likelihoods generated in ANGSD can be used to calculate statistics relevant to population structure analyses including inbreeding coefficients [13], F_{st} and principal component analyses (PCA) [14].

Another approach for detecting admixture including ancient admixture is the ABBA-BABA test also called the D-statistic [18]. For sequencing data the strategy for this test is based on sampling a single base at each position of the genome [24]. This strategy removes bias caused by depth differences which is a fundamental problem of NGS data. Given an outgroup ANGSD gives D-statistics for all possible combinations of the chosen individuals.

Association

Finding disease causing mutation is often done using association studies based on called genotypes. ANGSD provides two approaches for performing association studies that are appropriate for NGS data. Both are based directly on genotype likelihoods which takes all the uncertainty of the NGS data into account. The first method can be used in a simple case/control setting [10] where differences in allele frequencies between cases and controls are compared. The other approach is a more flexible generalized linear regression framework [11] which allows for quantitative traits and inclusion of covariates. This approach is also implemented for genotype probabilities such as the ones estimated from haplotype imputation.

Base Error Estimation

Several error estimates of type specific base error rates are implemented. The simplest is based on the mismatch rate that also forms the basis of SOAPsnp [6]. Another approach that tries to estimate the real error rate and not the mismatch rate is based on an outgroup [20],[24] and a high quality individual. The third approach estimates error rates, genotype likelihoods and allele frequencies simultaneously in order to determine the base error rate of polymorphic sites [10].

Limitations & Roadmap

Most statistical methods in ANGSD assume a diploid organism and does not support pooled data. Indels are represented internally in ANGSD, but no method currently utilizes this information. We also acknowledge that bcf/vcf files are heavily used and have begun including basic vcf input/output in the development version. No analysis in ANGSD uses pedigree information such as GATK's PhaseByTransmission. The CRAM format has been suggested as a successor to the BAM format, but ANGSD does not support this in the current version and depending on the general acceptance of this new format we might include it in future versions. Finally SAMtools and GATK include many different filters at the site level whereas these have not been included in ANGSD yet.

RESULTS AND DISCUSSION

ANGSD is the de facto(sole) implementation of many published methods (see Table 1), and we will in this section show examples of how to use ANGSD including a novel method of estimating the joint site frequency spectra for two populations and an implementation of the ABBA-BABA D-statistic [18] for NGS data. We will also show that having the ability to decide which method to use for some of the intermediate analyses, such as calculation of GL, is important and can have a large effect on the downstream analyses.

The Genotype Likelihood Model Affects Downstream Analysis

As an example of the effect of genotype likelihood model on the analysis, we estimated the SFS for 12 European (CEU) and 14 African (YRI) unrelated samples from the 1000 genomes project [25] sequenced using the Illumina platform. We used the method described in [9] implemented in ANGSD to estimate the site-frequency spectrum. This is a two step procedure that first involves calculating the sample allele frequency likelihoods followed by a numerical optimization for finding the maximum likelihood estimate of the SFS. Ancestral sites were obtained from the PanTro2 genome from the multiz46way dataset sync://hgdownload.cse.ucsc.edu/goldenPath/hg19/multiz46way/maf (available from the UCSC browser), and the analysis was based on a 170 Mb region from chromosome 1 by limiting our analysis to the sites with high mappability and discarding telomeres and centromeres. The ANGSD command used was

```
./angsd -bam CEU.list -rf regions.txt
-doSaf 1 -out ceu.gl1 -anc ancestral.fa
-GL 1 -nThreads 20 -sites filters.txt
```

We use the BAM files for the European samples listed in the file "CEU.list", limit our analysis to the non-centromeric/telomeric regions defined in the file "regions.txt", estimate the sample allele frequencies likelihoods (-doSaf), define the output files (-out) with prefix ceu.gl1, and use the genotype likelihood model from SAMtools (-GL 1). In order to estimate the joint allele frequency with YRI at a later stage we restrict the printed output to sites that are also present in the African sample and specified in the filters.txt filters.txtfile.

We also repeat the above analysis using the YRI population, and repeat the analyses for both populations using the GATK genotype likelihood model [8] (-GL 2).

From the sample allele frequency likelihoods for each site we then estimate the SFS using the Expectation Maximization (EM) algorithm:

```
./realSFS ceu.gl1.saf 24 -P 20 >ceu.sfs
```

Here realSFS is the secondary program written in c++c++, and finds the optima of equation (5) in [9]. We supply the realSFS with the file containing the sample allele frequency likelihoods (ceu.gl1.saf) and tell the program that the file contains 24 chromosomes (12 diploid individuals) and it should try to use 20 computer cores.

A)

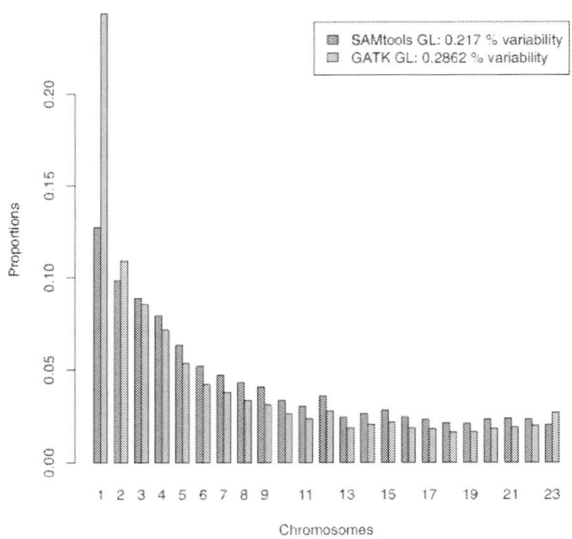

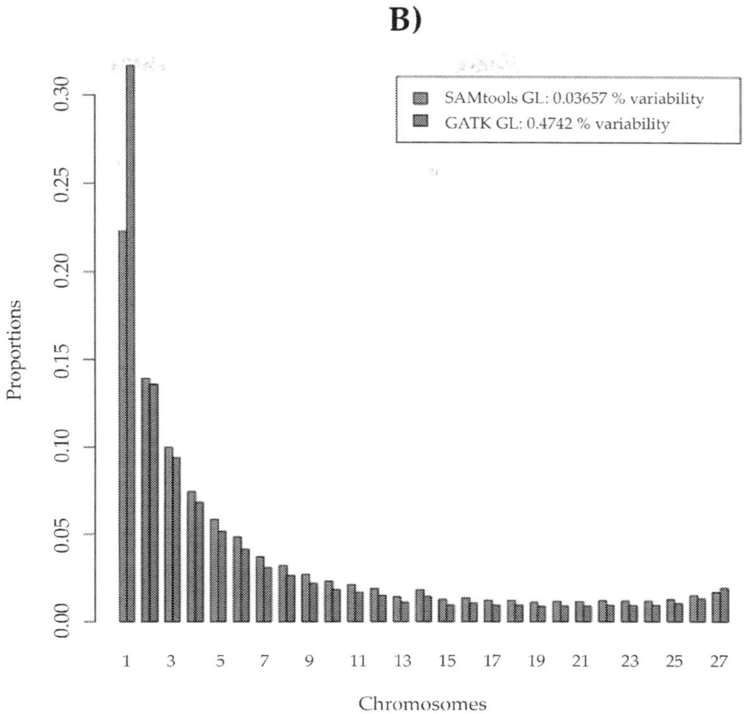

Figure 2: 1D SFS for different GL models. SFS estimation based on a 170 megabase region from chromosome 1 using 12 CEU samples **A)** and 14 YRI samples **B)**" from the 1000 genomes project. The analysis was performed for both the GATK GL model (green, light brown) and SAMtools GL (yellow,dark brown). Notice the difference in estimated variability (proportion of variable sites) for the two GL models, with GATK GL based analyses inferring more variable sites and an associated larger proportion of low-frequency alleles. The two categories of invariable sites have been removed and the distributions have been normalized so that the frequencies of all categories sum to one for each method.

The resulting four frequency spectra (SAMtools/GATK,CEU/YRI) are shown in Figure 2. From the figure it is evident that the analysis is highly sensitive to the chosen GL model. We emphasize that there is a clear need for more research on comparing methods for estimating GLs, and possible for developing new and more appropriate methods for estimating GLs. However, such research is beyond the scope of this paper. We here emphasize that the ANGSD approach for estimating the

SFS has been shown by others [26] to be superior to the genotype calling approaches used by SAMtools and GATK.

Joint Site Frequency Spectrum

We have generalized the approach for estimating the one dimensional SFS [9] to allow for two populations (see Methods section). To obtain the maximum likelihood estimate of the joint frequency spectrum we use an EM algorithm (equation 1) by evoking the following command:

```
./realSFS 2dsfs ceu.gl2.saf yri.gl2.saf
24 28 -P 30 >ceu.yri.sfs.
```

The result is shown on Figure 3. Unlike joint SFS based on SNP chip data (e.g. [27]), where most SNPs are polymorphic in both African and Europeans, this plot shows that most derived alleles are private to one of the populations. This is also observed between Chinese and Africans [28] and the difference between the SNP chip data and the sequencing data is caused by ascertainment biases in the chip data where SNPs are often chosen because they are common in populations such as European [27]. We have also performed a proper simulation study by simulating genotypes for two populations that follows a demographic pattern similar to European and African populations, assuming realistic recombination and mutation rates for humans. We simulated genotypes corresponding to a 10 Mb region using MSMS [29], and based on the genotypes we calculated genotype likelihoods using the method described in [12]. This was done by assuming a mean sequencing depth of 2X and an error rate of 0.2%. The true spectrum is visualized as a heat map in Additional file 1: Figure S1, and our estimated spectrum in Additional file 2: Figure S2 and Additional file 3: Figure S3.

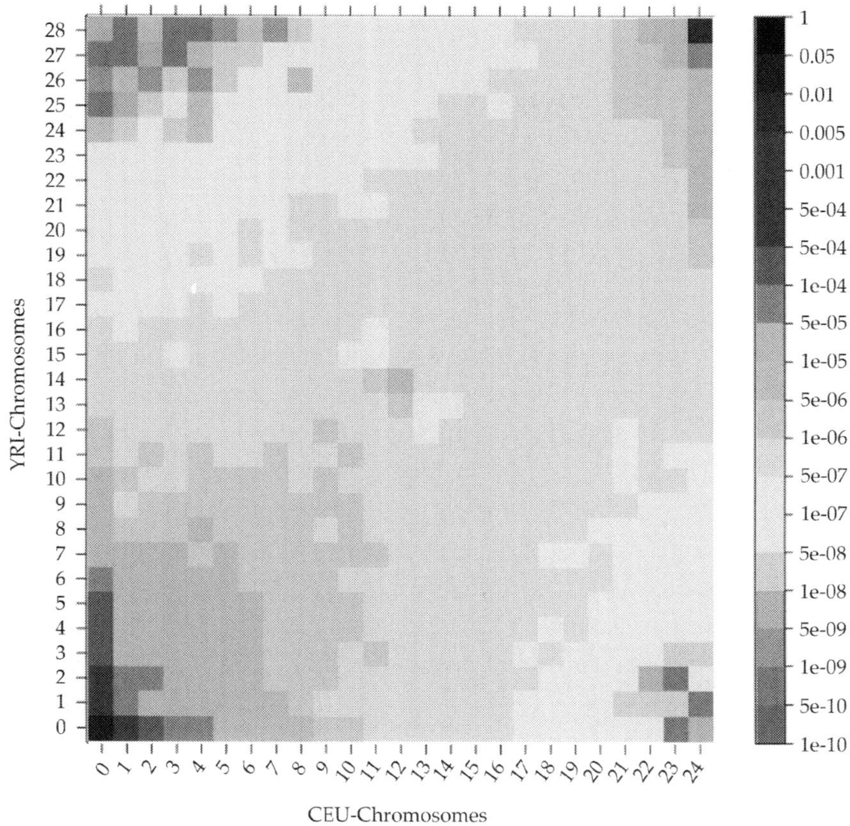

Figure 3: Joint SFS (2D-SFS). Two dimensional SFS estimation based on a 170 megabase region from chromosome 1 using 12 CEU samples and 14 YRI samples from the 1000 genomes project.

ABBA-BABA

To illustrate the use of the ABBA-BABA analyses [18],[24] we demonstrate two analyses: (1) an analysis of modern human samples, and (2) a comparison of modern human sequences and ancient DNA from the Denisovan hominin. For the modern individuals we tested a European (French), a Native American (Karitiana), a Papuan (Papuan1), a Han Chinese and an African (Yoruba) [30]. The ANGSD command used in the first analysis was

```
./angsd -doAbbababa 1 -bam modern.ind
-out modern -doCounts 1 -anc
chimpHG19.fa -minMapQ 30 -minQ 30
-blockSize 5000000
Rscript R/jackKnife.R
file=modern.abbababa
indNames=modern.indnames
```

The ABBA-BABA test is based on a sample of counts of bases (-doCount 1), an outgroup (-anc), which in this case is the chimpanzee, 5 Mb block size (-blockSize), and a strict filtering of bases based on quality scores (-minQ 30) and mapping quality (-minMapQ 30). A small Rscript is used to perform a blocked (uneven m-delete) jack knife procedure to obtain standard deviations and resulting the Z-scores. The results are shown in Table 2. The results are consistent with the current understanding of human migration out of Africa and even shows the recently proven link between Europeans and Native Americans through a shared ancestral central Asian population [31].

Table 2: D-stat results for modern samples

	H1	H2	H3	nABBA	nBABA	Dstat	jackEst	SE	Z
1	HGDP00521 (French)	HGDP00998 (American)	HGDP00927 (Yoruba)	355539	360029	-0.01	-0.01	0.00	-1.40
2	HGDP00521 (French)	HGDP00778 (Han china)	HGDP00927 (Yoruba)	361594	369006	-0.01	-0.01	0.00	-2.40
3	HGDP00998 (American)	HGDP00778 (Han china)	HGDP00927 (Yoruba)	332227	334990	-0.00	-0.00	0.00	-0.90
4	HGDP00521 (French)	HGDP00542 (Papuan1)	HGDP00927 (Yoruba)	360153	383994	-0.03	-0.03	0.00	-6.80
5	HGDP00998 (American)	HGDP00542 (Papuan1)	HGDP00927 (Yoruba)	347593	366979	-0.03	-0.03	0.00	-5.80
6	HGDP00778 (Han china)	HGDP00542 (Papuan1)	HGDP00927 (Yoruba)	347017	363467	-0.02	-0.02	0.00	-5.20
7	HGDP00927 (Yoruba)	HGDP00998 (American)	HGDP00521 (French)	653515	360029	0.29	0.29	0.00	60.60
8	HGDP00927 (Yoruba)	HGDP00778 (Han china)	HGDP00521 (French)	639280	369006	0.27	0.27	0.01	53.00

9	HGDP00998 (American)	HGDP00778 (Han china)	HGDP00521 (French)	384915	407967	-0.03	-0.03	0.01	-5.40
10	HGDP00927 (Yoruba)	HGDP00542 (Papuan1)	HGDP00521 (French)	626366	383994	0.24	0.24	0.01	43.10
11	HGDP00998 (American)	HGDP00542 (Papuan1)	HGDP00521 (French)	399343	450303	-0.06	-0.06	0.01	-10.10
12	HGDP00778 (Han china)	HGDP00542 (Papuan1)	HGDP00521 (French)	405942	433790	-0.03	-0.03	0.01	-5.50
13	HGDP00927 (Yoruba)	HGDP00521 (French)	HGDP00998 (American)	653515	355539	0.30	0.30	0.00	61.20
14	HGDP00927 (Yoruba)	HGDP00778 (Han china)	HGDP00998 (American)	711281	334990	0.36	0.36	0.01	71.80
15	HGDP00521 (French)	HGDP00778 (Han china)	HGDP00998 (American)	486385	407967	0.09	0.09	0.01	15.10
16	HGDP00927 (Yoruba)	HGDP00542 (Papuan1)	HGDP00998 (American)	660154	366979	0.29	0.29	0.01	53.80
17	HGDP00521 (French)	HGDP00542 (Papuan1)	HGDP00998 (American)	445929	450303	-0.00	-0.00	0.01	-0.80
18	HGDP00778 (Han china)	HGDP00542 (Papuan1)	HGDP00998 (American)	394958	477720	-0.09	-0.09	0.01	-15.30
19	HGDP00927 (Yoruba)	HGDP00521 (French)	HGDP00778 (Han china)	639280	361594	0.28	0.28	0.00	57.00
20	HGDP00927 (Yoruba)	HGDP00998 (American)	HGDP00778 (Han china)	711281	332227	0.36	0.36	0.01	72.70
21	HGDP00521 (French)	HGDP00998 (American)	HGDP00778 (Han china)	486385	384915	0.12	0.12	0.01	20.80
22	HGDP00927 (Yoruba)	HGDP00542 (Papuan1)	HGDP00778 (Han china)	666222	363467	0.29	0.29	0.01	55.10
23	HGDP00521 (French)	HGDP00542 (Papuan1)	HGDP00778 (Han china)	459135	433790	0.03	0.03	0.01	4.70
24	HGDP00998 (American)	HGDP00542 (Papuan1)	HGDP00778 (Han china)	401357	477720	-0.09	-0.09	0.01	-14.20
25	HGDP00927 (Yoruba)	HGDP00521 (French)	HGDP00542 (Papuan1)	626366	360153	0.27	0.27	0.01	54.00
26	HGDP00927 (Yoruba)	HGDP00998 (American)	HGDP00542 (Papuan1)	660154	347593	0.31	0.31	0.01	60.60
27	HGDP00521 (French)	HGDP00998 (American)	HGDP00542 (Papuan1)	445929	399343	0.06	0.06	0.01	9.50

28	HGDP00927 (Yoruba)	HGDP00778 (Han china)	HGDP00542 (Papuan1)	666222	347017	0.32	0.32	0.01	61.90
29	HGDP00521 (French)	HGDP00778 (Han china)	HGDP00542 (Papuan1)	459135	405942	0.06	0.06	0.01	10.40
30	HGDP00998 (American)	HGDP00778 (Han china)	HGDP00542 (Papuan1)	401357	394958	0.01	0.01	0.01	1.30

Results of the ABBABABA analysis for modern individuals from the human genetic diversity panel.

In the second analysis we used the following commands:

```
./angsd -doAbbababa 1 -bam denisova.ind
-out deniRes -doCounts 1 -anc
chimpHG19.fa -minMapQ 30 -minQ 30
-rmTrans 1

Rscript R/jackKnife.R
file=deniRes.abbababa
indNames=denisona.indnames
```

In the command line above, we removed transitions (-rmTrans) which are known to have extremely high error rates for ancient genomes. A more elaborate scheme for filtering bases using base quality scores can also be used to specify a different threshold for each individual and each of the four bases, and has also been implemented [24],[32]. The results for the tests are shown in Table 3. This test for introgression between Papuan ancestors and Denisovans rejects the tree (((Yoruban,Papuan),Denisova), chimpanzee), with a Z score of 12.1, in accordance with the current understanding in the field [30],[32].

Table 3: D-stat for ancient sample

	H1	H2	H3	nABBA	nBABA	Dstat	jackEst	SE	Z
1	HGDP00927 (Yoruba)	HGDP00542 (Papuan1)	T_hg19_1000g (Denisova)	103016	90667	0.06	0.06	0.01	12.10
2	T_ hg19_1000g (Denisova)	HGDP00542 (Papuan1)	HGDP00927 (Yoruba)	286551	90667	0.52	0.52	0.00	127.10
3	T_ hg19_1000g (Denisova)	HGDP00927 (Yoruba)	HGDP00542 (Papuan1)	286551	103016	0.47	0.47	0.01	88.60

Results of the ABBABABA analysis for 2 modern individuals and one ancient sample.

SNP Discovery and Genotype Calling

Population genetic analyses are traditionally based on called genotypes, but this poses a significant problem for NGS data due to the nature of the technology. Genotypes are not directly observable, but must be inferred from the data. For low or medium coverage data there can be considerable uncertainty in genotype inferences, potentially leading to errors or biases in downstream analyses. Arguably, the optimal solution to this problem is to avoid genotype calling altogether, and instead base inferences on methods that incorporate genotype uncertainty with the GLs [9]-[14],[22],[26],[33]. However, we recognize that many analyses have not been generalized to be based on GLs instead of called genotypes, and we have therefore included basic SNP discovery and genotype calling into ANGSD, using methods that efficiently can take advantage of estimated priors derived from GL based analyses. In ANGSD SNPs are inferred based on allele frequency estimation using a likelihood ratio test that can reject that the allele frequency is 0 [10]. We compare SNP calling using GATK (UnifiedGenotyper, default parameters), SAMtools (-q 10) and ANGSD based on 33 CEU samples from the 1000 genomes project [25]. ANGSD, SAMtools and GATK take into account the quality of the called bases (qscores) by modeling the uncertainty of possible genotypes, but differ in GL model, SNP calling criterion, filtering, etc (see [34]). In the commands below we perform SNP calling for all combinations of 1) p-value of site being variable 10^{-6},0.01**-snp_pval** -snp_pval 2) using local qscore recalibration with the BAQ model [35] -baq -baq 3) SAMtools or GATK GL model **-GL** -GL.

```
./angsd -b bam.list -doMaf 1 -doMajorMinor
 1 -snp_pval 0.01 -GL 1 -P 4
 -baq 0 -ref hg19.fa -minQ 13 -minMapQ 10
```

Venn diagrams of the overlap of sites are shown in Figure 4 (p-value $<10^{-6}$, no BAQ), Additional file 4: Figure S4 (p-value $<10^{-2}$, no BAQ), Additional file 5: Figure S5 (p-value $<10^{-2}$, with BAQ), and Additional file 6: Figure S6 (p-value $<10^{-6}$, with BAQ). Notice the difference in variable sites for the different GL models, and the decrease of variable sites when applying BAQ. When choosing a lenient p-value threshold (0.01) ANGSD infers more SNP sites than the other two methods when choosing a strict p-value threshold (10^{-6}) fewer sites are called. In these analyses we removed sites that were called as homozygous for alternative alleles for SAMTools and GATK.

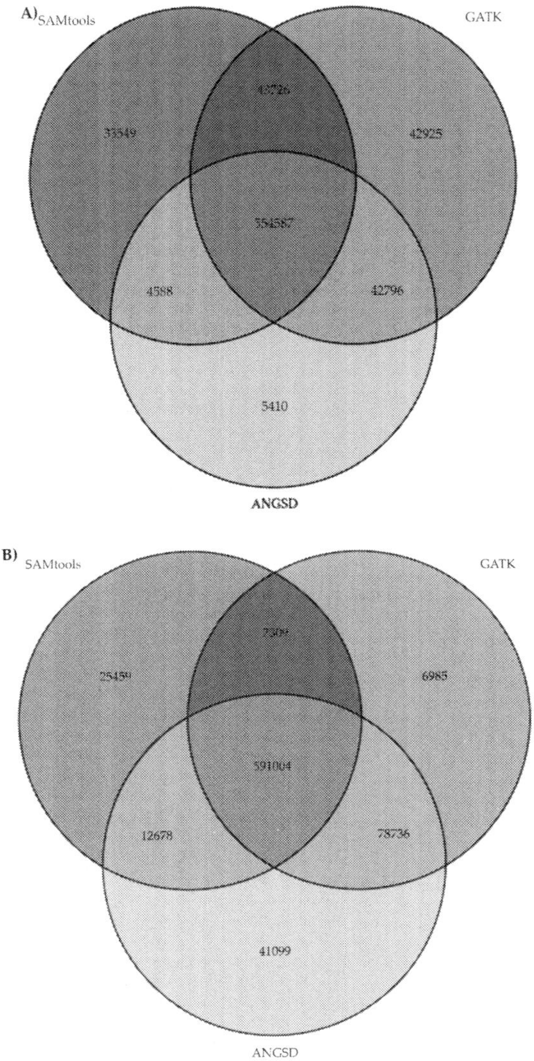

Figure 4: Overlap between inferred SNPs with a critical p-value threshold of 10 $^{-6}$ and not using BAQ. Venn diagram of the overlap between the SNP discovery for ANGSD, GATK and SAMtools for 33 CEU samples for chromosome 1. We used default parameters with GATK for SAMtools we discarded reads with a mapping quality below 10. For ANGSD we choose an p-value threshold of 10^{-6} and didn't enable BAQ. In **A**, we used the SAMtools genotype likelihood model in ANGSD, in **B** we used the GATK model in ANGSD.

Assuming a segregating site is diallelic, there are 3 possible genotypes. In ANGSD we allow for 3 different methods for calculating genotype posteriors (GP), and we can define simple genotype calling criteria using these posteriors. We can either choose the genotype with the maximum posterior probability, or more sensibly, we can define a cutoff such that a genotype will be set to missing if it is below a certain threshold. Our 3 models for calculating GP are 1) assuming uniform prior (raw ML based on GL) (ML) 2) using an estimate of the population frequency as prior (AF) [10] 3) using the SFS as prior by calculating the genotype probabilities for an individual conditional on the information for all individuals [9] (SFS). We compared the three different ANGSD approaches with two existing tools for genotype calling (SAMtools,GATK) by using 31 CEU individuals that are part of the 1000genomes project and the HapMap project [36]. The exact commands used for this analysis in Additional file 7.

We include reference genome information for all methods even though ANGSD does not need the information. Additionally, we force all methods to call genotypes for all sites. Each genotype call is assigned a probability or quality score. A threshold can then be applied to remove low quality calls. For sites where the different method did not provide a genotype call we set the genotype as homozygous for the reference allele and give the call the worst possible quality score. The results for the 1,456,587 HapMap sites for all 5 methods are shown in Figure 5. The jaggedness of the SAMtools/GATK curves are due to the discretization of the phred scaled genotype qualities. We observe no big difference between the different methods for high call rates. For lower call rates we see that the ML method in ANGSD is somewhere between the GATK and SAMTools methods. For very low call rates we see that SAMtools outperforms the other methods.

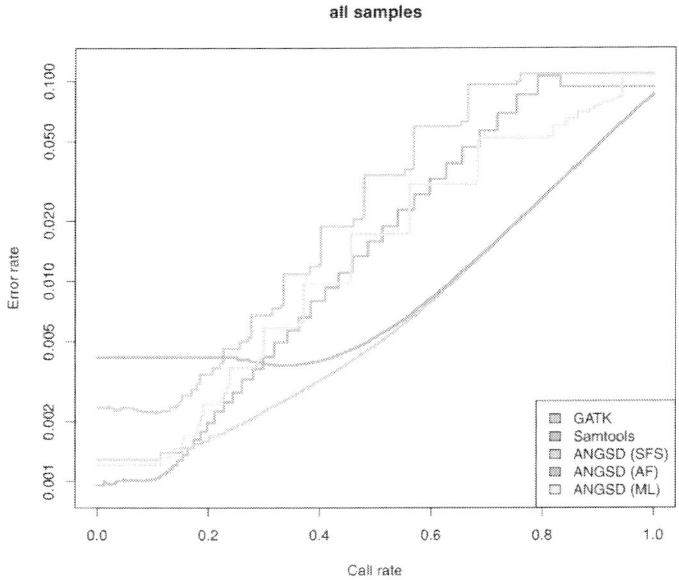

Figure 5: Error rate vs call rate for called genotypes. Error rate and call rates for genotype calls based on different methods. The error rate is defined as the discordance rate between HapMap genotype calls compared to the same individuals sequenced in the 1000 genomes. Genotype where called for all sites for all individuals for all methods. Each genotype call has a score which was used to determine the call rate. Due to the discrete nature of some of the genotype scores we obtain a jagged curve.

Computational Speed

To compare the running time of ANGSD with existing tools we performed simple SNP discovery and allele frequency estimation for ANGSD, GATK and SAMtools. This we did with 50 samples, 100 samples and 200 samples (using the first 20 megabase region of human chromosome 21). For ANGSD and GATK we redid the analysis, but this time allowing for 2 and 4 cores (commands used are found in Additional file 7). The result is shown in Table 4. A fair comparison between tools is impossible for several reasons: 1) The tools perform slightly different analyses. 2) The storage subsystem might behave very different on different systems. 3) SAMtools is in its current version non thread-able in downstream analyses. 4) Difference in workload at runtime. In practice most analyses can be run in parallel over different chromosomes or even smaller regions, which makes the lack of threading in SAMtools more of a nuisance than a real problem. We would expect the runtime to be linear in the region size for all programs, and observed similar memory usage for all three tools. From the table we observe that ANGSD is faster in all scenarios, but we emphasize that cannot conclude that ANGSD is generally faster, but we do not expect that it is significantly slower than SAMtools and GATK on any given system. We also note that the bottleneck for simple analyses is the file reading, and allocating an unlimited number of cores will not decrease the running time, but might actually increase it. This is what we observe for both ANGSD and GATK for this simple analysis. In a more complex analysis setting such as SAF estimation we would expect a better utilization of the threads. See Additional file 7 for commands used.

Table 4: Computational speed of GATK,SAMtools and ANGSD

	SAMtools	GATK	ANGSD	GATK (2cores)	ANGSD (2cores)	GATK (4cores)	ANGSD (4cores)
50 Samples	2722	1706	602	1744	1171	1765	1646
100 Samples	5097	4049	1270	4143	2457	4373	4013
200 Samples	10615	9672	2704	9951	5032	10330	7352

Wallclock time (not CPU) measured in seconds for different samples sizes and different number of allocated cores. Commands used are found in Additional file 7. We did the analysis twice (in different order) and picked the lowest value. Notice that the runtime for GATK and ANGSD does not decrease with 2 and 4 threads. This could be an indication that the file reading is the bottleneck.

CONCLUSIONS

We have developed a fast program for analyses of NGS data that enable researchers to perform various analyses, particularly population genetic analyses that are not implemented in any other existing programs. For many of the analyses we use the full information of the data by avoiding genotype and SNP calling and instead basing analyses on GLs, calculated using different methods, typically using quality scores. This is especially useful for low-coverage data and for non-human organisms where imputation cannot be performed reliably due to the lack of a reference population.

Availability and requirements

- Project name: ANGSD (version 0.612 or higher)
- Project home page: http://www.popgen.dk/angsd, https://github.com/ANGSD.
- Operating system(s): Platform independent. But only tested on the Linux distribution Ubuntu.
- Programming language: c/c++.
- Other requirements: zlib. For some downstream analysis R R is required.
- License: GPL version 2.
- Any restrictions to use by non-academics: None.

ACKNOWLEDGEMENTS

We acknowledge the authors of SOAPsnp and SAMtools for using open source code, Line Skotte and Ida Moltke for helpful discussions, and Mateo Fumagalli and Filipe Vieira for their development of analyses tools for ANGSD. Work for this manuscript was financed through Danish National Research Foundation (DNRF94), the Villum Foundation, and grant 2R14003229-07 from the National Institutes of Health.

ADDITIONAL FILES

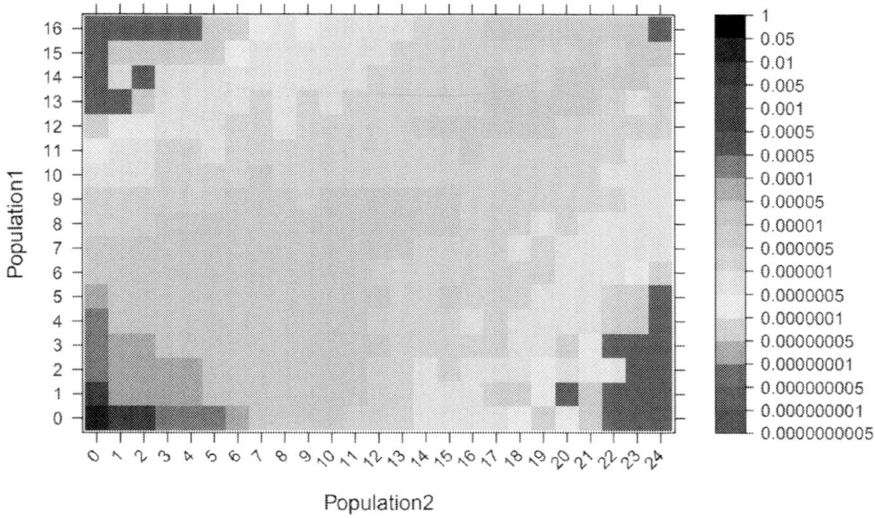

Additional file 1: Figure S1. True 2D site frequency spectrum. A heatmap of the two dimensional site frequency spectrum simulated on the basis of known genotypes using a demographic model to mimick 12 European individuals and 8 african samples. The estimated spectrum can be found in Additional file 2: Figure S2.

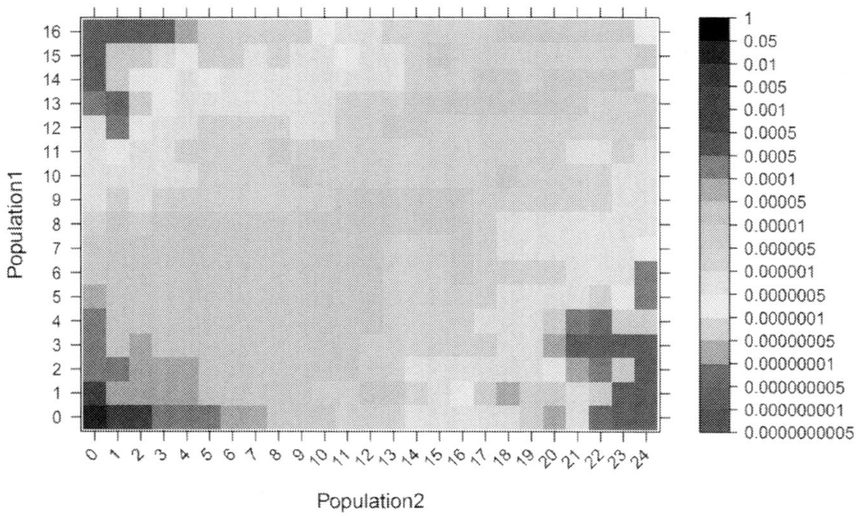

Additional file 2: Figure S2. Estimated 2D site frequency spectrum. A heatmap of the two dimensional site frequency spectrum estimated on the basis of genotype likelihoods for simulated genotypes. Data was simulated assuming a sequencing depth of 2X and an errorrate of 0.2%. The true estimates are seen in Additional file 1: Figure S1, and the difference between the true and the estimated can be found in Additional file 3: Figure S3.

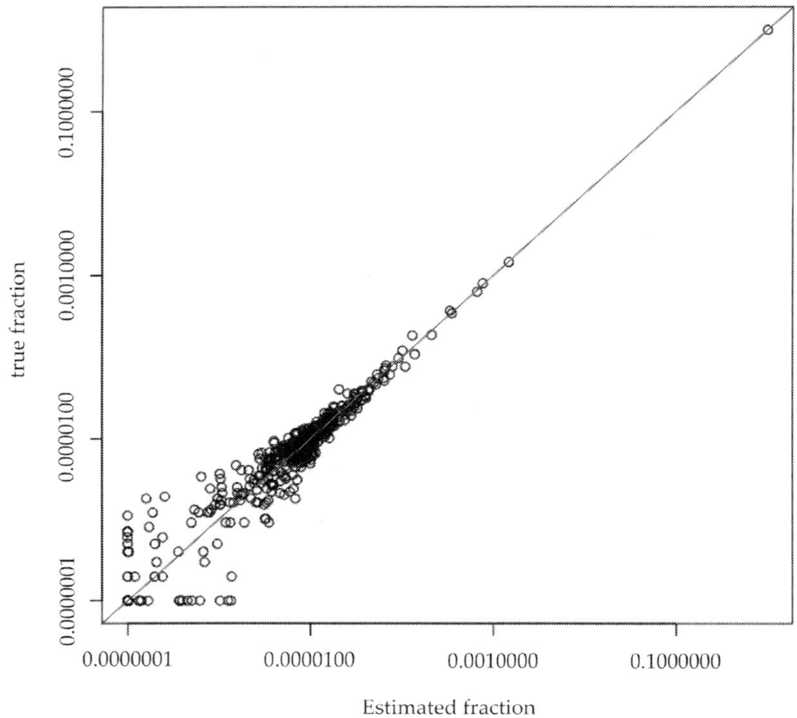

Additional file 3: Figure S3. Difference between true proportions vs the estimated proportions. Plot of the estimated proportions and the true proportions. The estimated proportions are based on genotype likelihoods calculated assuming 2X sequencing depth and 0.2% error rate. The genotypes are simulated using msms and should reflect the difference European individuals (bottleneck followed by rapid expansion), and African individuals.

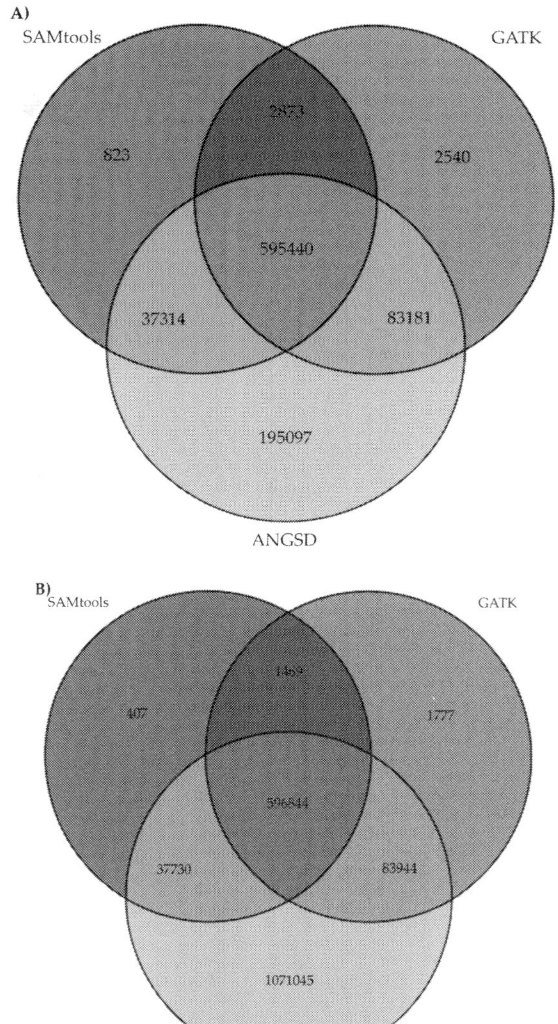

Additional file 4: Figure S4. Overlap between inferred SNPs, a critical p-value threshold of 10^{-2} and not using BAQ. Venn diagram of the overlap between the SNP discovery for ANGSD, GATK and SAMtools for 33 CEU samples for chromosome 1. We used default parameters with GATK for SAMtools we discarded reads with a mapping quality below 10. For ANGSD we choose an p-value threshold of 0.01 and didn't enable BAQ. In A, we used the SAMtools genotype likelihood model in ANGSD, in B we used the original GATK GL model in ANGSD.

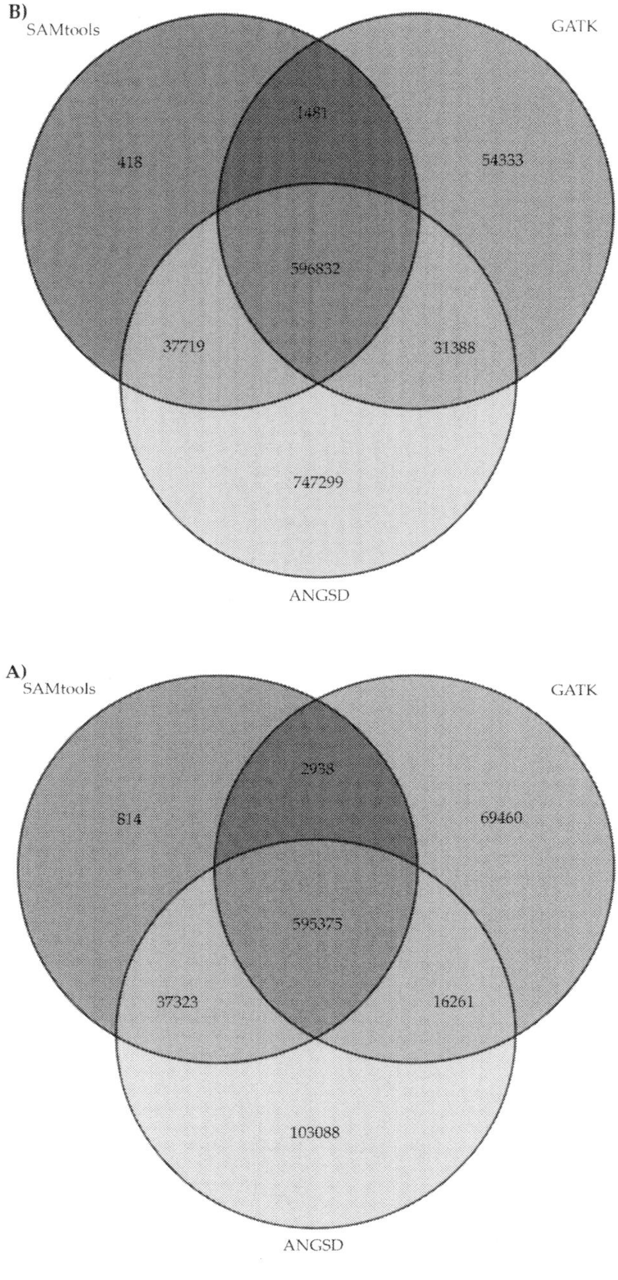

Additional file 5: Figure S5. Overlap between inferred SNPs, a critical p-value threshold of 10^{-2} with BAQ. Venn diagram of the overlap between the SNP discovery for ANGSD, GATK and SAMtools for 33 CEU samples for chromosome 1. We used default parameters with GATK for SAMtools we discarded reads with a mapping quality below 10. For ANGSD we choose a p-value threshold of 0.01 and enabled BAQ. In A, we used the SAMtools genotype likelihood model in ANGSD, in B we used the GATK model in ANGSD.

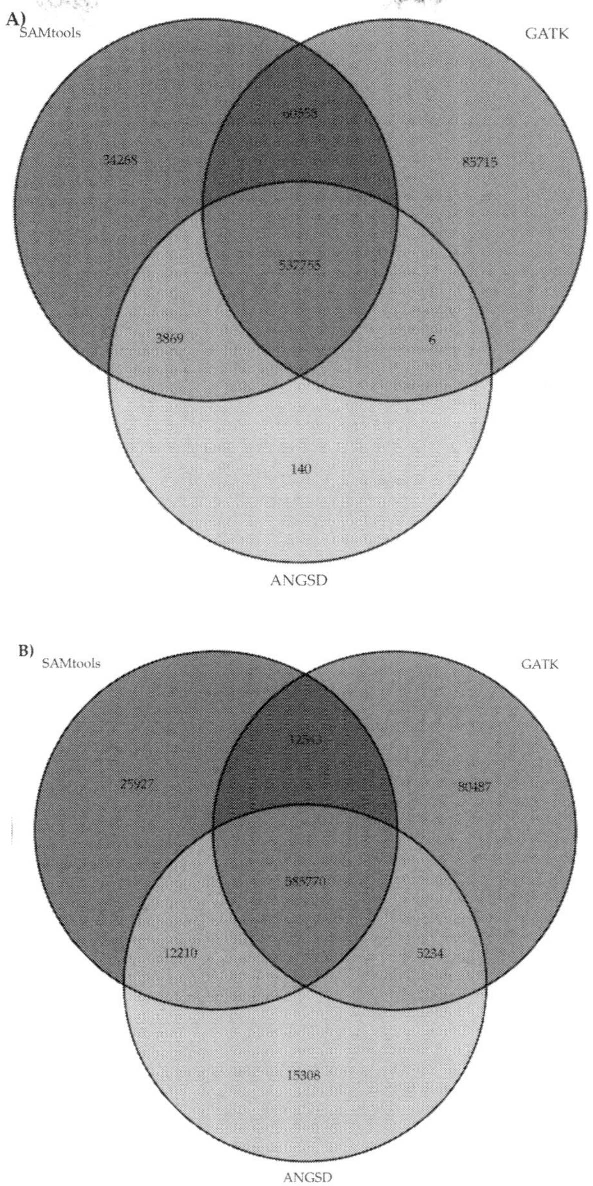

Additional file 6: Figure S6. Overlap between SNP sites, a critical value of 10^{-6} with BAQ. Venn diagram of the overlap between the SNP discovery for ANGSD, GATK and SAMtools for 33 CEU samples for chromosome 1. We used default parameters with GATK for SAMtools we discarded reads with a mapping quality below 10. For ANGSD we choose a p-value threshold of 10^{-6} and enabled BAQ. In A, we used the SAMtools genotype likelihood model in ANGSD, in B we used the GATK model in ANGSD.

```
## genotype calling
#ANGSD AF
angsd -sites sites.txt -b ceu31.list -doMajorMinor 4 -doMaf 1 -GL 1 -doGeno 19 -doPost 1 -ref
hg19.fa
#ANGSD ML
angsd -sites sites.txt -b ceu31.list -doMajorMinor 4 -GL 1 -doGeno 19 -doPost 2 -ref hg19.fa
#ANGSD SFS
angsd -sites sites.txt -b ceu31.list -doMajorMinor 4 -GL 1 -doSaf 3 -P 6 -ref  hg19.fa -pest
sfs.txt
#SAMtools
./samtools mpileup -f  hg19.fa -b ceu31.list -g -I -l sites.txt |bcftools view -cg  -
#GATK
./gatk -l INFO -R hg19.fa -T HaplotypeCaller -I ceu31.list -L sites.intervals --output_mode
EMIT_ALL_SITES
## computational speed
samtools mpileup -b file.list -f hg19.fa -g -r 21:1-20000000 |bcftools view -v
gatk -R hg19.fa -T UnifiedGenotyper -I file.list -L 21:1-20000000 -nt 1
angsd -ref hg19.fa -b file.list -doMaf 1 -GL 1 -P 1 -doMajorMinor 1 -r 21:1-20000000 -
snp_pval 1e-5 -checkbamheaders 0
```

Additional file 7: Commands used for some of the analyses. Text file containing the commands used in various analysis in the text. We used SAMtools version 0.1.19-44428cd, and GATK version 2.4-7-g5e89f01.

AUTHORS' CONTRIBUTIONS

TSK and AA wrote the code for ANGSD, with input from RN. TSK coded the core algorithms for data parsing and most data analysis algorithms. TSK and AA performed the data analyses in this paper. All authors contributed to the writing of the manuscript, with TSK writing the first draft. All authors read and approved the final manuscript.

REFERENCES

1. Nielsen R, Paul JS, Albrechtsen A, Song YS: Genotype and SNP calling from next-generation sequencing data . Nat Rev Genet. 2011, 12 (6): 443-451. 10.1038/nrg2986.
2. Li H, Durbin R: Fast and accurate short read alignment with Burrows-Wheeler transform . Bioinformatics. 2009, 25: 1754-1760. 10.1093/bioinformatics/btp324.
3. Li R, Yu C, Li Y, Lam TW, Yiu SM, Kristiansen K, Wang J: SOAP2:

an improved ultrafast tool for short read alignment . Bioinformatics. 2009, 25: 1966-1967. 10.1093/bioinformatics/btp336.

4. Langmead B, Trapnell C, Pop M, Salzberg SL: Ultrafast and memory-efficient alignment of short DNA sequences to the human genome . Genome Biol. 2009, 10 (3): R25-10.1186/gb-2009-10-3-r25.
5. Marco-Sola S, Sammeth M, Guigo R, Ribeca P: The GEM mapper: fast, accurate and versatile alignment by filtration . Nat Methods. 2012, 9 (12): 1185-1188. 10.1038/nmeth.2221.
6. Li R, Li Y, Fang X, Yang H, Wang J, Kristiansen K, Wang J: SNP detection for massively parallel whole-genome resequencing . Genome Res. 2009, 19 (6): 1124-1132. 10.1101/gr.088013.108.
7. Cabanski CR, Cavin K, Bizon C, Wilkerson MD, Parker JS, Wilhelmsen KC, Perou CM, Marron JS, Hayes DN: ReQON: a Bioconductor package for recalibrating quality scores from next-generation sequencing data . BMC Bioinformatics. 2012, 13: 221-10.1186/1471-2105-13-221.
8. McKenna A, Hanna M, Banks E, Sivachenko A, Cibulskis K, Kernytsky A, Garimella K, Altshuler D, Gabriel S, Daly M, DePristo MA: The Genome Analysis Toolkit: a MapReduce framework for analyzing next-generation DNA sequencing data . Genome Res. 2010, 20 (9): 1297-1303. 10.1101/gr.107524.110.
9. Nielsen R, Korneliussen T, Albrechtsen A, Li Y, Wang J: SNP calling, genotype calling, and sample allele frequency estimation from new-generation sequencing data . PLoS ONE. 2012, 7 (7): e37558-10.1371/journal.pone.0037558.
10. Kim SY, Lohmueller KE, Albrechtsen A, Li Y, Korneliussen T, Tian G, Grarup N, Jiang T, Andersen G, Witte D, Jorgensen T, Hansen T, Pedersen O, Wang J, Nielsen R: Estimation of allele frequency and association mapping using next-generation sequencing data . BMC Bioinformatics. 2011, 12: 231-10.1186/1471-2105-12-231.
11. Skotte L, Korneliussen TS, Albrechtsen A: Association testing for next-generation sequencing data using score statistics . Genet Epidemiol. 2012, 36 (5): 430-437. 10.1002/gepi.21636.
12. Korneliussen T, Moltke I, Albrechtsen A, Nielsen R: Calculation of Tajima's D and other neutrality test statistics from low depth next-

generation sequencing data . BMC Bioinformatics. 2013, 14: 289-10.1186/1471-2105-14-289. [http://www.biomedcentral.com/1471-2105/14/289],

13. Vieira FG, Fumagalli M, Albrechtsen A, Nielsen R: Estimating inbreeding coefficients from NGS data: Impact on genotype calling and allele frequency estimation . Genome Res. 2013, 23 (11): 1852-1861. 10.1101/gr.157388.113.
14. Fumagalli M, Vieira FG, Korneliussen TS, Linderoth T, Huerta-Sanchez E, Albrechtsen A, Nielsen R: Quantifying population genetic differentiation from next-generation sequencing data . Genetics. 2013, 195 (3): 979-992. 10.1534/genetics.113.154740.
15. Li H, Durbin R: Fast and accurate short read alignment with Burrows-Wheeler transform . Bioinformatics. 2009, 25: 1754-1760. 10.1093/bioinformatics/btp324.
16. Browning BL, Yu Z: Simultaneous genotype calling and haplotype phasing improves genotype accuracy and reduces false-positive associations for genome-wide association studies . Am J Hum Genet. 2009, 85 (6): 847-861. 10.1016/j.ajhg.2009.11.004.
17. Fumagalli M, Vieira FG, Linderoth T, Nielsen R: ngsTools: methods for population genetics analyses from next-generation sequencing data . Bioinformatics. 2014, 30 (10): 1486-1487. 10.1093/bioinformatics/btu041.
18. Patterson N, Moorjani P, Luo Y, Mallick S, Rohland N, Zhan Y, Genschoreck T, Webster T, Reich D: Ancient admixture in human history . Genetics. 2012, 192 (3): 1065-1093. 10.1534/genetics.112.145037.
19. Rasmussen M, Li Y, Lindgreen S, Pedersen JS, Albrechtsen A, Moltke I, Metspalu M, Metspalu E, Kivisild T, Gupta R, Bertalan M, Nielsen K, Gilbert MT, Wang Y, Raghavan M, Campos PF, Kamp HM, Wilson AS, Gledhill A, Tridico S, Bunce M, Lorenzen ED, Binladen J, Guo X, Zhao J, Zhang X, Zhang H, Li Z, Chen M, Orlando L, et al: Ancient human genome sequence of an extinct Palaeo-Eskimo . Nature. 2010, 463 (7282): 757-762. 10.1038/nature08835.
20. Orlando L, Ginolhac A, Zhang G, Froese D, Albrechtsen A, Stiller M, Schubert M, Cappellini E, Petersen B, Moltke I, Johnson PL,

Fumagalli M, Vilstrup JT, Raghavan M, Korneliussen T, Malaspinas AS, Vogt J, Szklarczyk D, Kelstrup CD, Vinther J, Dolocan A, Stenderup J, Velazquez AM, Cahill J, Rasmussen M, Wang X, Min J, Zazula GD, Seguin-Orlando A, Mortensen C, et al: Recalibrating Equus evolution using the genome sequence of an early Middle Pleistocene horse . Nature. 2013, 499 (7456): 74-78. 10.1038/nature12323.

21. Li Y, Vinckenbosch N, Tian G, Huerta-Sanchez E, Jiang T, Jiang H, Albrechtsen A, Andersen G, Cao H, Korneliussen T, Grarup N, Guo Y, Hellman I, Jin X, Li Q, Liu J, Liu X, Sparsø T, Tang M, Wu H, Wu R, Yu C, Zheng H, Astrup A, Bolund L, Holmkvist J, Jørgensen T, Kristiansen K, Schmitz O, Schwartz TW, et al: Resequencing of 200 human exomes identifies an excess of low-frequency non-synonymous coding variants . Nat Genet. 2010, 42 (11): 969-972. 10.1038/ng.680.

22. Skotte L, Korneliussen TS, Albrechtsen A: Estimating individual admixture proportions from next generation sequencing data . Genetics. 2013, 195 (3): 693-702. 10.1534/genetics.113.154138.

23. Li H: A statistical framework for SNP calling, mutation discovery, association mapping and population genetical parameter estimation from sequencing data . Bioinformatics. 2011, 27 (21): 2987-2993. 10.1093/bioinformatics/btr509.

24. Green RE, Krause J, Briggs AW, Maricic T, Stenzel U, Kircher M, Patterson N, Li H, Zhai W, Fritz MH, Hansen NF, Durand EY, Malaspinas AS, Jensen JD, Marques-Bonet T, Alkan C, Prufer K, Meyer M, Burbano HA, Good JM, Schultz R, Aximu-Petri A, Butthof A, Hober B, Hoffner B, Siegemund M, Weihmann A, Nusbaum C, Lander ES, Russ C, et al: A draft sequence of the Neandertal genome . Science. 2010, 328: 710-722. 10.1126/science.1188021.

25. Consortium TGP: A map of human genome variation from population-scale sequencing . Nature. 2010, 467 (7319): 1061-1073. 10.1038/nature09534. [http://www.nature.com/nature/journal/v467/n7319/full/nature09534.html],

26. Han E, Sinsheimer JS, Novembre J: Characterizing bias in population genetic inferences from low-coverage sequencing data . Mol Biol Evol. 2014, 31 (3): 723-735. 10.1093/molbev/mst229.

27. Albrechtsen A, Nielsen FC, Nielsen R: Ascertainment biases in SNP chips affect measures of population divergence . Mol Biol Evol. 2010, 27 (11): 2534-2547. 10.1093/molbev/msq148.
28. Gravel S, Henn BM, Gutenkunst RN, Indap AR, Marth GT, Clark AG, Yu F, Gibbs RA, Bustamante CD, Altshuler DL, Durbin RM, Abecasis GR, Bentley DR, Chakravarti A, Clark AG, Collins FS, De la Vega FM, Donnelly P, Egholm M, Flicek P, Gabriel SB, Gibbs RA, Knoppers BM, Lander ES, Lehrach H, Mardis ER, McVean GA, Nickerson DA, Peltonen L, Schafer AJ, et al: Demographic history and rare allele sharing among human populations . Proc Natl Acad Sci U S A. 2011, 108 (29): 11983-11988. 10.1073/pnas.1019276108.
29. Ewing G, Hermisson J: MSMS: a coalescent simulation program including recombination, demographic structure and selection at a single locus . Bioinformatics. 2010, 26 (16): 2064-2065. 10.1093/bioinformatics/btq322.
30. Meyer M, Kircher M, Gansauge MT, Li H, Racimo F, Mallick S, Schraiber JG, Jay F, Prufer K, de Filippo C, Sudmant PH, Alkan C, Fu Q, Do R, Rohland N, Tandon A, Siebauer M, Green RE, Bryc K, Briggs AW, Stenzel U, Dabney J, Shendure J, Kitzman J, Hammer MF, Shunkov MV, Derevianko AP, Patterson N, Andres AM, Eichler EE, et al: A high-coverage genome sequence from an archaic Denisovan individual . Science. 2012, 338 (6104): 222-226. 10.1126/science.1224344.
31. Raghavan M, Skoglund P, Graf KE, Metspalu M, Albrechtsen A, Moltke I, Rasmussen S, Stafford TW, Orlando L, Metspalu E, Karmin M, Tambets K, Rootsi S, Magi R, Campos PF, Balanovska E, Balanovsky O, Khusnutdinova E, Litvinov S, Osipova LP, Fedorova SA, Voevoda MI, DeGiorgio M, Sicheritz-Ponten T, Brunak S, Demeshchenko S, Kivisild T, Villems R, Nielsen R, Jakobsson M, et al: Upper Palaeolithic Siberian genome reveals dual ancestry of Native Americans . Nature. 2014, 505 (7481): 87-91. 10.1038/nature12736.
32. Reich D, Green RE, Kircher M, Krause J, Patterson N, Durand EY, Viola B, Briggs AW, Stenzel U, Johnson PL, Maricic T, Good JM, Marques-Bonet T, Alkan C, Fu Q, Mallick S, Li H, Meyer M, Eichler EE, Stoneking M, Richards M, Talamo S, Shunkov MV,

Derevianko AP, Hublin JJ, Kelso J, Slatkin M, Paabo S: Genetic history of an archaic hominin group from Denisova Cave in Siberia . Nature. 2010, 468 (7327): 1053-1060. 10.1038/nature09710.

33. Wang Y, Lu J, Yu J, Gibbs RA, Yu F: An integrative variant analysis pipeline for accurate genotype/haplotype inference in population NGS data . Genome Res. 2013, 23 (5): 833-842. 10.1101/gr.146084.112.

34. Yu X: Sun S: Comparing a few SNP calling algorithms using low-coverage sequencing data . BMC Bioinformatics. 2013, 14: 274-10.1186/1471-2105-14-274.

35. Li H: Improving SNP discovery by base alignment quality . Bioinformatics. 2011, 27 (8): 1157-1158. 10.1093/bioinformatics/btr076.

36. Frazer KA, Ballinger DGEa: A second generation human haplotype map of over 3.1 million SNPs . Nature. 2007, 449 (7164): 851-861. 10.1038/nature06258.

CHAPTER 20

NEXT GENERATION SEQUENCING OF ANCIENT DNA: REQUIREMENTS, STRATEGIES AND PERSPECTIVES

Michael Knapp [1] and Michael Hofreiter [2]

[1]Allan Wilson Centre for Molecular Ecology and Evolution, Department of Anatomy and Structural Biology, University of Otago, 270 Great King Street, 9016 Dunedin, New Zealand

[2]Department of Biology, The University of York, Wentworth Way, Heslington, YO10 5DD, York, UK

ABSTRACT

The invention of next-generation-sequencing has revolutionized almost all fields of genetics, but few have profited from it as much as the field of ancient DNA research. From its beginnings as an interesting but rather marginal discipline, ancient DNA research is now on its way into the center of evolutionary biology. In less than a year from its invention next-generation-sequencing had increased the amount of DNA sequence data available from extinct organisms by several orders of magnitude. Ancient DNA research is now not only adding a temporal aspect to evolutionary studies and allowing for the observation of evolution in real time, it also provides important data to help understand the origins of our own species. Here we review progress that has been made in next-generation-sequencing of ancient DNA over the past five years and evaluate sequencing strategies and future directions.

Keywords: Ancient DNA, barcoding, genome sequencing, hybridization capture, targeted sequencing

INTRODUCTION

The first study reporting ancient DNA (aDNA) sequences in 1984, predating the invention of PCR, was on 221 base pairs (bp) of mitochondrial DNA (mtDNA) obtained from an approximately 140 years old quagga skin [1]. In the late 1980s the invention of PCR [2,3,4] provided a significant boost to aDNA studies, including a much increased time depth of up to several tens of thousands of years. Nevertheless, until recently, analyses of aDNA were mostly restricted to short DNA fragments, usually derived from the mitochondrial genome, which is present in living cells in far higher copy numbers than the nuclear genome. Recent improvements in PCR technology have helped overcome some of the problems restricting ancient DNA research to short mtDNA fragments, but no innovation has revolutionized aDNA research to the extent that the emergence of next-generation-sequencing (NGS) has done. The first of these technologies was described in 2005 [5] and was almost immediately implemented in aDNA research. Within a few month of the introduction of NGS Poinar *et al.* [6] published 13 million bp from the nuclear genome of the extinct woolly mammoth. When compared with the 27,000 bp of cave bear sequence [7] that represented the largest nuclear data set available from an extinct species in the pre-NGS era, the data set obtained by Poinar *et al.* represented a 480x increase. This development has so far peaked in the publication of low coverage draft nuclear genomes of the woolly mammoth (0.8-fold [8]) and the Neanderthal (1-fold [9]) and also a high-quality 20-fold coverage nuclear genome of a 4,500 year old Palaeo-Eskimo [10]. The second, somewhat more recent technological change that is starting to influence aDNA research to an increasing extent is the move from PCR to hybridization capture as a method for enriching for the desired target sequence. For modern DNA, hybridization capture has successfully been used to enrich for large parts of the genome, for example the complete exome [11]. It is now starting to replace PCR as the method of choice for targeting specific sequences in the sea of contaminating environmental DNA typical for most aDNA samples [12,13,14,15].

In this review, we provide an overview of the strengths and weaknesses of the various next generation sequencing strategies used to sequence aDNA and the ways hybridization capture can be used to enrich for the sequence of interest, even from multiple samples. Finally, we discuss which directions aDNA research may take in the future.

TECHNICAL ISSUES

Sequencing Instruments and Features

The four commonly used instruments for massively parallel sequencing are Roche (454) GS FLX Titanium, Illumina (Solexa) Genome Analyzer IIe, Applied Biosystems SOLiD and Helicos BioSciences HeliScope. A good overview is available in Millar *et al.* [16], although capacity and read length have increased since the publication of this review. The Titanium update of the GS FLX now produces about 400-600 million base pairs (megabases – MB) per instrument run with read lengths of up to 400 bp while the Illumina Solexa Genome Analyzer IIe produces up to 48 gigabases (GB) with read length up to two times 100 bp in paired end reads. Ancient DNA is usually highly fragmented with average fragment lengths ranging from 51.3 bp for some Neanderthal DNA [12] to 142 and 164 bp, respectively, for DNA from permafrost mammoth hair for which the largest DNA fraction was gel-purified [8]. Thus both the 454 and the Illumina read length is sufficient for sequencing aDNA fragments across their full length. Both platforms provide the possibility to physically separate lanes on a sequencing plate, thereby allowing for sequencing of multiple libraries without the necessity of barcoding. While a 454 picotiter plate can be separated into up to 16 individual lanes, the Illumina genome analyzer allows for separation into up to eight lanes. No studies using either the SOLiD or the HeliScope platform for aDNA sequencing have so far been published; therefore we will not discuss them in this review. Given the rapid speed of innovation in the field of next-generation-sequencing, it does appear likely, though, that both the SOLiD and HeliScope platform will be utilized for aDNA sequencing in the near future.

Selection of Instrument

Apart from availability, the choice of instrument is mainly influenced by the desired application and the quality of DNA used.

The latest model of the 454 sequencer provides average read lengths of more than 400 bp. This fragment length is sufficient to greatly reduce potential errors in *de novo* assemblies of consecutive sequence. However, of all next generation sequencing instruments it produces the smallest amount of data from a single run. This disadvantage is even greater if the average molecule length of the target is less than 400 bp, which is generally the case for aDNA, at least when the full range of aDNA fragments is targeted, be it via shot-gun sequencing or when using hybridization capture as enrichment technology. The reason for the on average relatively short fragment length of endogenous aDNA lies in the fact that the copy number of aDNA molecules increases by a factor of 2 -100 when the fragment length is divided by two [6,17]. However, when PCR is used to amplify relatively long fragments above 200 bp, which is often possible for permafrost samples (e.g. [18,19,20]), then 454 can be the most sensible choice.

All other next generation sequencing instruments produce shorter read length but substantially more sequence data. Given the low amount of endogenous DNA and the short molecule length characteristic for aDNA, the larger amount of sequence data produced by these machines is a clear advantage for most aDNA studies. This is especially true for genome sequencing projects where the lower costs per nucleotide dramatically reduce both the overall price and the time required for a project [10], but also for studies using hybridization capture as usually the enriched sequences are not pure but are still embedded in a fairly high background of 60 – 80 % non-specific sequences [12,15].

Library Preparation

The preparation of sequencing libraries is very similar for both 454 and Illumina sequencing. First, molecules of ideal length for the respective sequencing approach are produced either by shearing longer molecules or by PCR amplification of ideally sized target fragments. As aDNA is characterized by strongly fragmented DNA molecules, shearing is usually not necessary, and may in fact be detrimental [21]. DNA sequencing libraries are in both cases constructed by ligation of universal adapters to both ends of target molecules. These sequencing adapters contain priming sites for sequencing and amplification.

While both Illumina and 454 library preparation protocols are suitable for library preparation from modern DNA and other high copy number templates such as PCR products, they have limitations when dealing with low copy number aDNA extracts. The 454 library preparation protocol requires a streptavidin bead purification of adapter ligated template. A recent study [22] found that the last step, in which the non-biotinylated DNA strand of each molecule is released from the streptavidin capture beads by incubation with NaOH, is highly inefficient. In fact, only 0.1% of adaptor ligated library bound to the streptavidin beads was eluted from the beads using this approach. The authors found that replacing the NaOH elution step by short heat treatment increased the yield to 98%.

The Illumina library preparation protocol is not ideal for aDNA either. The cohesive-end ligation of adapters described in the Illumina protocol requires one more purification step compared with the 454 blunt end ligation. It could be shown [22] that in each Qiagen Minelute purification between 16% and 40% of the template is lost, which makes each additional purification step detrimental with regard to the final yield of the library. Furthermore, the Illumina library preparation includes a pre-amplification gel purification step instead of the 454 streptavidin bead purification, which is not ideal for aDNA analyses. As well as being a potential source for the introduction of contamination, it may result in the loss of significant amounts of template.

Due to these shortcomings, new strategies to improve the efficiency and workflow of the library preparation process have been developed. Roche have recently introduced a rapid protocol to produce target fragments with adenine overhangs suitable for cohesive-end ligation of

454 sequencing adapters that does not require an extra purification step compared with the blunt ending procedure. Strategies for improving the efficiency of the Illumina libary preparation have also been developed [23]. These improvements in library preparation have dramatically reduced the amount of template DNA required. Therefore, for most applications, including shotgun sequencing even of less well preserved samples, library preparation is no longer a limiting factor in NGS analyses of aDNA.

NEXT GENERATION SEQUENCING OF ANCIENT DNA

More than 20 studies have already made use of NGS to obtain sequence data from ancient remains (Table 1). Different sequencing strategies have been applied depending on the sample material and the objective of the study.

Shotgun Sequencing

Shotgun sequencing of aDNA can provide valuable information about the quality of the extract. It allows analyzing the composition of a DNA extract and to determine the relative abundance of endogenous DNA compared with contaminants. Ancient DNA extracts have been found to have very different levels of contamination. Sequencing 28 MB of DNA from a 28,000 year old, permafrost preserved mammoth from Siberia, Poinar *et al.* [6] found that at least 13 MB (45.4%) could be identified as being derived from the mammoth genome. Thus it takes only about twice as much sequencing effort to obtain the entire genome from this mammoth sample than it would to sequence for example a modern elephant genome. For sequencing the mammoth draft genome, a hair sample containing more than 90 % endogenous DNA was chosen [8] and a similarly high content of endogenous DNA was found for the hair tuft of a 4,500 year old Palaeo-Eskimo [10]. Preservation like this is however rare and can usually only be found in very young samples or in remains that have been preserved under optimal conditions such as permafrost. Subfossil remains of similar age preserved under temperate conditions such as Neanderthals and cave bears frequently contain less than 10% of endogenous DNA [7,9,24]. A further quality measure that can be determined from shotgun sequencing of aDNA is the average fragment length of the endogenous DNA. Fragment length hardly ever exceeds 400 bp and can be much shorter. Briggs *et al.* [12] found the average mitochondrial fragment length in five Neanderthal extracts to be between 51.3 and 79.3 bp. This information can be used to determine further sequencing strategies.

Shotgun sequencing represents the simplest sequencing strategy using NGS. It has been used for sequencing complete genomes of extinct species from well preserved permafrost mammoths [8,10] and from more poorly preserved Neanderthal remains [9,24] and sequencing of mitochondrial genomes from numerous species [36,37]. While a complete mammalian genome is about 2.8 - 4 billion bp, the mitochondrial genome is only about 17,000 bp long. Thus, the mitochondrial genome makes up only about 0.0004% of the entire genome. However, due to high copy numbers per cell, reads covering mitochondrial genome sequences are proportionally overrepresented among shotgun reads. In a recent study Gilbert *et al.* [25] found that, depending on the sample, between 0.08% and 1.99% of the reads obtained by shotgun sequencing mammoth extracts represented mitochondrial sequences. Thus, even the 454 sequencer, which produces only about 25,000 – 40,000 reads in its smallest run unit (1/16 of a full picotiter plate) will produce up to 800 mitochondrial reads. With a read length of 400 bp this would result in 18-fold coverage of the mitochondrial genome from each of the 16 lanes on a 454 picotiter plate, but even with an average read length of 100 bp the mitochondrial genome would still be covered 4-fold.

However, this strategy is only suitable for very well preserved samples. Sequencing libraries from poorly preserved Neanderthal remains can contain less than 0.03% mitochondrial reads [28], as a much larger part of the overall reads is taken up by bacterial and other environmental DNA. The approach is also unsuitable for well preserved ancient samples if any other genomic region is targeted, as almost all other regions except for the mitochondrial genome are present in only 1-2 copies per cell. And even if the aim is to sequence mitochondrial genomes from well preserved samples, a success rate of 2% means that 98% of the reads are not useful for the analyses. Therefore, most aDNA studies require target enrichment strategies, which may soon also become the method of choice for complete genome sequencing, as they drastically reduce sequencing costs and have, in the form of hybridization capture, already been used to obtain high coverage DNA sequence data for ~14,000 protein coding positions of the nuclear genome from poorly preserved Neanderthal DNA [15].

Target Enrichment Strategies

Two main categories of target enrichment techniques commonly employed in aDNA research can be distinguished: 1) Polymerase chain reaction (PCR) amplification of target regions and 2) DNA capture via hybridization.

Table 1. Ancient DNA studies using next generation sequencing partly or exclusively for sequence data collection

Author/Date	Publication Title	Target region	Sequencing/ Target enrichment Strategy
Poinar *et al.* Jan. 2006 [6]	Metagenomics to paleogenomics: large-scale sequencing of mammoth DNA	genome	shotgun
Green *et al.* Nov. 2006 [24]	Analysis of one million base pairs of Neanderthal DNA	genome	shotgun
Gilbert *et al.* Sep 2007 [25]	Whole-genome shotgun sequencing of mitochondria from ancient hair shafts	mitochondrial genome	shotgun
Gilbert *et al.* Jun. 2008 [26]	Intraspecific phylogenetic analysis of Siberian woolly mammoths using complete mitochondrial genomes	mitochondrial genome	shotgun
Gilbert *et al.* Jun. 2008 [27]	Paleo-Eskimo mtDNA genome reveals matrilineal discontinuity in Greenland	mitochondrial genome	shotgun

Green *et al.* Aug. 2008 [28]	A complete Neandertal mitochondrial genome sequence determined by high-throughput sequencing	mitochondrial genome	shotgun
Miller *et al.* Nov. 2008 [8]	Sequencing the nuclear genome of the extinct woolly mammoth	genome	shotgun
Miller *et al.* Feb. 2009 [29]	The mitochondrial genome sequence of the Tasmanian tiger (Thylacinus cynocephalus)	mitochondrial genome	shotgun
Allentoft *et al.* Mar. 2009 [30]	Identification of microsatellites from an extinct moa species using high-throughput (454) sequence data	microsatellites	shotgun
Ramírez *et al.* May 2009 [31]	Paleogenomics in a temperate environment: shotgun sequencing from an extinct Mediterranean caprine	genome	shotgun
Willerslev *et al.* May 2009 [32]	Analysis of complete mitochondrial genomes from extinct and extant rhinoceroses reveals lack of phylogenetic resolution	mitochondrial genome	shotgun
Briggs *et al.* Jul. 2009 [12]	Targeted retrieval and analysis of five Neandertal mtDNA genomes	mitochondrial genome	capture

Zhao, Qi and Schuster Aug. 2009 [33]	Tracking the past: interspersed repeats in an extinct Afrotherian mammal, Mammuthus primigenius	genome	shotgun
Stiller *et al.* Oct. 2009 [34]	Direct multiplex sequencing (DMPS)--a novel method for targeted high-throughput sequencing of ancient and highly degraded DNA	mitochondrial genome	multiplex PCR
Krause *et al.* Feb. 2010 [13]	A complete mtDNA genome of an early modern human from Kostenki, Russia	mitochondrial genome	capture
Rasmussen *et al.* Feb. 2010 [10]	Ancient human genome sequence of an extinct Palaeo-Eskimo	genome	shotgun
Edwards *et al.* Feb. 2010 [35]	A complete mitochondrial genome sequence from a mesolithic wild aurochs (Bos primigenius)	mitochondrial genome	shotgun
Lindqvist *et al.* Mar. 2010 [36]	Complete mitochondrial genome of a Pleistocene jawbone unveils the origin of polar bear	mitochondrial genome	shotgun
Krause *et al.* Apr. 2010 [14]	The complete mitochondrial DNA genome of an unknown hominin from southern Siberia	mitochondrial genome	capture

Burbano *et al.* May 2010 [15]	Targeted investigation of the Neandertal genome by array-based sequence capture	nuclear genome	capture
Green *et al.* May 2010 [9]	A draft sequence of the Neandertal genome	genome	shotgun

PCR

For both classical Sanger sequencing [38] and NGS approaches, PCR is still the most commonly used target enrichment strategy in aDNA research, although this situation may change in the near future.

PCR target enrichment for NGS of modern DNA normally involves long range amplification of the target region followed by shearing of the PCR products to a size suitable for the NGS instrument used [39]. Ancient DNA however is usually highly degraded and not suitable for long range PCR. Therefore, if longer genomic regions such as the complete mitochondrial genome are to be sequenced, numerous independent PCR amplifications of overlapping short fragments are necessary. As aDNA extract is in many cases available in limited supply, it can be the limiting factor that determines how many PCR amplifications are possible. In 2005, a multiplex PCR approach that partially ameliorates this problem was introduced into ancient DNA research [20]. Instead of amplifying each amplicon separately, a number of non-overlapping amplicons are amplified in the same reaction, using the same amount of template as a singleplex PCR would have required. Using this approach, as little as two-non overlapping multiplex PCR may be sufficient to amplify long consecutive stretches of target DNA [20,40,41,42,43]. The PCR products from these first step multiplex amplifications can then be used in different ways.

In pre-NGS studies using multiplex PCR in connection with Sanger sequencing [20], the multiplex product was used as template for further second step singleplex amplifications. These second step amplifications were then used as template for Sanger sequencing. This approach can be adapted for NGS. All second step singleplex amplicons can be pooled, transformed into a sequencing library by ligation of the sequencing adapters required by the respective NGS instrument and then sequenced. The advantage of this approach is that singleplex amplicons can be quantified individually and pooled in equimolar ratios to obtain close to equal distribution of sequencing reads across all amplicons. The disadvantage is that singleplex amplification can be very time consuming if long target regions from highly degraded samples are to be amplified via a large number of very short amplicons.

As an alternative to the singleplex amplification using the multiplex product as template, the multiplex template itself can be directly transformed into a sequencing library without further singleplex amplification [34]. This is a very time efficient approach, but it has the disadvantage that unequal efficiencies of different primer pairs in a multiplex PCR will manifest themselves in unequal sequencing coverage of different amplicons. This effect can be limited by optimizing cycle numbers in the multiplex PCR (enough to lift the target above the background but not too many to prevent too much variation in the copy number between poorly and well amplifying amplicons). Although such optimization requires additional sequencing and thus produces additional costs, this approach is to date the most time efficient PCR enrichment method for long target regions from degraded samples.

Despite their status as the exclusive means of target enrichment in aDNA studies for about 20 years, PCR approaches have some significant shortcomings when used on aDNA. First and foremost, PCR requires a certain target molecule length to produce meaningful results. About 30 bp of informative sequence are needed to assign any given molecule confidently to its accurate coordinates within the genome [44]. Thus, if each PCR primer is 20 bp long, molecules of about 70 bp length are required to obtain 30 bp of informative sequence. As a consequence, many previous studies have targeted amplicons of 100 bp and more. However, 70 bp is already at the upper limit of average fragment length identified for many ancient extracts [7,10,12]. Even for permafrost preserved mammoth samples, average fragment lengths of more than 100 bp were only obtained by selecting the largest fragment fraction via gel electrophoresis [8]. Therefore, in almost all cases, PCR will only target a tiny fraction of the molecules at the very top end of the fragment length distribution, thereby greatly reducing information contained in the extract and increasing the risk of starting amplifications from single molecules.

Especially in studies of humans and Neanderthals this effect can cause severe contamination of the resulting sequences. As modern human DNA can be an abundant source of contamination and modern human molecules will exceed 70 bp in length more often than the ancient molecules, all PCR approaches enrich for contaminating modern human molecules. Therefore, even small overall amounts of contamination can lead to erroneous sequences. In contrast, enrichment via hybridization capture such as PEC [12] targets the more abundant short endogenous aDNA molecules. This way, the ratio of endogenous to contaminating hominid DNA can be shifted substantially in favor of the former. This allows for authentic consensus sequences to be obtained even from samples that are relatively strongly contaminated when judged by PCR [13]. Finally, standard PCR (in contrast to quantitative PCR, which can however, not be applied to the multiplex approaches discussed above) does not reveal any information about the number of molecules a PCR started from. Therefore, at least two independent PCRs are required to ensure that not all reads covering any given site are amplicons of the same, potentially damaged or misread starting molecule [45]. Most of these problems of PCR discussed above do not occur in hybridization capture approaches.

Nevertheless, it remains to be tested whether hybridization capture can outperform PCR enrichment if complex mixtures of species are targeted. In studies on sediment or ice core DNA [46,47,48], the use of conserved primers that flank variable regions has worked well. Given that hybridization capture relies on similarity between probes and target molecules, it may be difficult to develop capture systems that allow for the enrichment of variable regions across species. Probes targeting conserved regions adjacent to variable regions may represent a viable, albeit an as yet untested approach.

Hybridization Capture

Numerous hybridization capture methods have been introduced for use with modern DNA. Strategies include on-array capture as well as in-solution capture [49,50,51,52,53]. In all cases target DNA is hybridized with specific probes that are either immobilized on microarrays (Figure 1) or can be immobilized on specific capture beads.

So far, hybridization capture has been used in aDNA research to a very limited extent. The first study, to our knowledge, using a hybridization approach for targeting specific aDNA sequences was published in 2006 when Noonan *et al.* [54] showed that it was possible to capture specific, even nuclear, DNA targets from a batch culture of a Neanderthal library cloned into bacteria. They used biotinylated oligos that are, after an in-solution hybridization step, bound to streptavidin beads and targeted loci which they knew from sequencing several thousand clones from the bacterial library, should be present in the library. A different approach was taken by Anderung *et al.* [55], who used biotinylated oligos for capturing short fragments of mitochondrial DNA directly from aDNA extracts. This principle was taken up in 2009 by Briggs *et al.* [12,56] who used it to enrich mitochondrial DNA from five 454 Neanderthal sequencing libraries.

In all cases, the principal is very similar to commercial in-solution capture strategies. Target specific, biotinylated oligos are hybridized to the genomic regions that are to be analyzed. The library/probe hybrid is then immobilized on magnetic, streptavidin covered beads. Non-target library is removed by washing and the target library is subsequently eluted from the streptavidin beads. The method is suitable for any NGS instrument, but it requires library preparation from aDNA extracts. Thus, the library preparation has to be optimized for very low copy numbers as described above. Moreover, this approach generally requires library amplification before hybridization to obtain a suitable amount of template and re-amplification of the enriched library after hybridization to ensure a sufficient DNA concentration for sequencing.

Despite these additional amplification steps, hybridization capture as a target enrichment approach for aDNA overcomes some of the limitations of PCR enrichment. It requires a much shorter minimal fragment length than PCR and can therefore draw on a much bigger pool of endogenous, highly degraded molecules instead of just targeting the few long molecules from the upper limit of the fragment length distribution. This effect not only increases the potential to retain most of the original complexity of a sequencing library, it also reduces the risk of enriching modern contamination, as described above [13]. Finally, as individual molecules are sequenced and can be distinguished by their start and end coordinates with respect to the reference sequence, hybridization capture also allows determining the sequence coverage of each position in the same way as in shotgun sequencing.

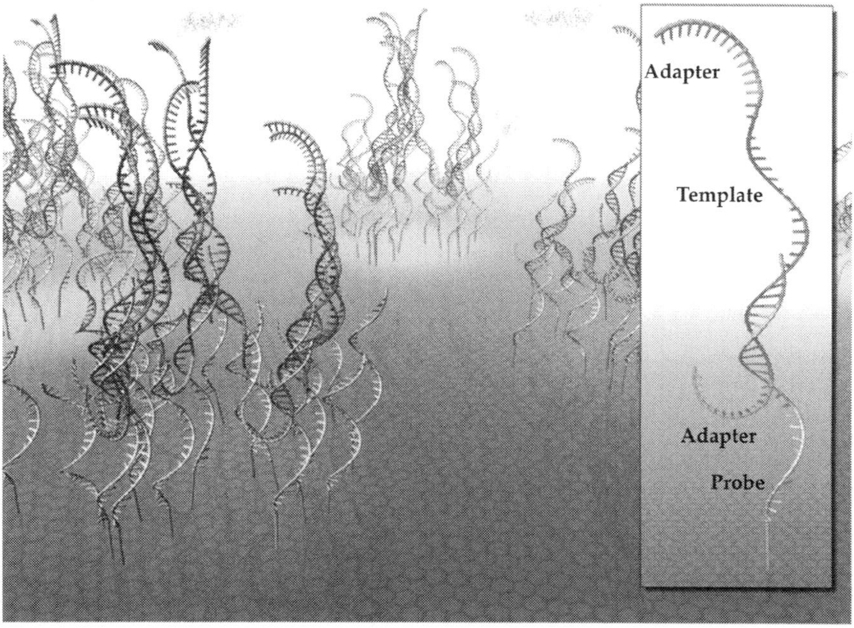

Figure 1. Hybridization capture on a microarray. Single-stranded oligonucleotide probes are arranged in clusters (red, yellow, green and blue surface colors) on a glass slide. Captured sequencing libraries consist of single-stranded target molecules (blue) and flanking sequencing adapters (green).

The main disadvantage of hybridization capture compared with PCR enrichment is the loss of template molecules during library preparation. It remains to be studied whether the potential to capture shorter molecules balances for this loss. In general it can be assumed that the shorter the average fragment length, the more efficient hybridization capture is compared to any PCR approach.

Barcoding Sequencing Libraries

For many studies, for example in population genetics, even the smallest lane on any NGS instrument will produce excessive amounts of sequence data for a single sample. Moreover, if larger numbers of samples are to be analyzed, the costs soon become prohibitive if a full single lane is used per sample. It is therefore important to be able to pool multiple samples and sequence them in a single lane. As the information about sample origin of individual sequence reads is lost in all NGS approaches, barcoding techniques that put a specific tag on all DNA fragments from each respective sample and thereby allow reads to be assigned to samples using bioinformatics approaches are often necessary. Several different strategies are commonly used for barcoding sequencing libraries.

Roche Fusion Primers

Roche offers a PCR based solution to transform PCR amplicons into barcoded or non-barcoded 454 sequencing libraries. Amplicons are amplified with oligos containing an amplicon specific priming site, an overhang containing the 454 adapter sequence and key and if desired a multiplex identifier (MID) between key and amplicon specific priming site (Figure 2 (a)). The approach is simple, quick and efficient, but it also has two major disadvantages. First, it requires an individual barcoded primer set for each individual that is included in the sequencing pool. This can rapidly increase costs and is therefore not feasible for population level studies. Second, the approach is unsuitable for library preparation directly from DNA extracts or sheared PCR products.

Parallel Tagged Sequencing

Parallel tagged sequencing (PTS) is a ligation based barcoding protocol [39,57]. In this approach, palindromic barcoding oligos are ligated to both ends of double-stranded target molecules to produce blunt ended, barcoded target molecules ready for library preparation (Figure 2 (b)). PTS can in principle be used with any NGS instrument. However, as the first sequenced nucleotide (i.e the first nucleotide 3' of the sequencing adapter) is identical for all molecules in this approach, it is not ideal for Illumina sequencing, as this will prevent the calling of sequence clusters. PTS is particularly suitable for the barcoding of any pre-amplified template. Templates can be barcoded individually and subsequently quantified to allow barcoded templates to be pooled in equimolar ratios prior to library preparation. However, similar to the fusion primer approach, the barcode is invariably located 3' of the sequencing primer and thus read as part of the sequencing read. It therefore effectively reduces the length of the target read by the length of the barcode (commonly about 7 – 10 nucleotides). While this is not an issue with the 400 bp reads of the

454 instrument, it is a limitation to be considered when the approach is used with instruments that produce shorter reads. A further disadvantage for aDNA studies is that the approach requires two subsequent adapter ligations. As each one is associated with a significant amount of template loss, the approach is not suitable for low copy number starting templates such as aDNA extracts or low cycle PCR products.

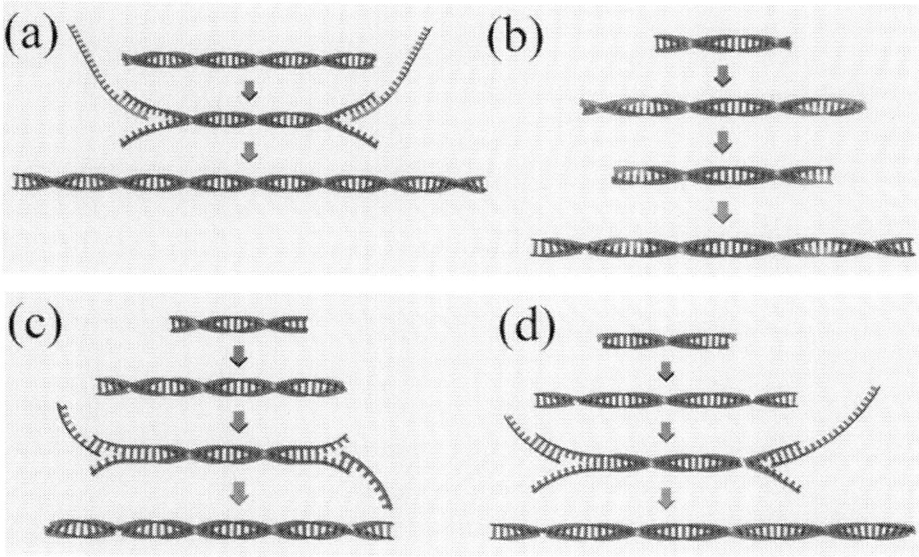

Figure 2. Comparison of barcoding strategies. **(a)** Fusion primers (454): Target molecules (blue) are amplified with primers consisting of target specific priming site (green), barcode (red) and 454 specific adapter sequence (grey). **(b)** Parallel tagged sequencing (PTS): Palindromic barcodes (red/green) are blunt end ligated to the ends of target molecules (blue). The outer half of the barcoding adapter (*i.e.*, the reverse complement part) is removed by restriction digest. Sequencing adapters (grey) are blunt end ligated to the ends of the barcoding adapters. **(c)** Direct multiplex sequencing (DMPS): One truncated, barcoded adapter (red and grey) and one truncated standard adapter (grey) are blunt end ligated to the ends of target molecules (blue). Adapters are extended to full length by PCR amplification with sequencing adapter specific primers. **(d)** Illumina barcoding (simplified): Truncated Illumina adapters (grey) are cohesive-end ligated to template molecules (blue) with adenine overhangs (green). Adapters are extended to full length by PCR amplification with sequencing adapter specific primers. One of the primers contains the barcoding sequence (red).

Direct Multiplex Sequencing

Direct multiplex sequencing (DMPS) [34] was developed to facilitate barcoding of multiplex PCR enriched aDNA and allows for omitting the time consuming singleplex step. It follows very similar principles as PTS, but instead of blunt end ligating barcoding adapters independent from the standard library preparation adapters, DMPS allows for ligation of truncated, barcoded library preparation adapters followed by PCR amplification using primers that extend the adaptors to full length (Figure 2 (c)). It therefore requires one adapter ligation step less than PTS. In DMPS, multiplex PCRs are usually conducted with a minimum number of PCR cycles, to prevent well performing primer pairs in the multiplex set from outperforming less well performing primer pairs [34]. As a result of the small number of starting molecules in aDNA and the low number of PCR cycles, multiplex PCR products themselves usually have very small copy numbers, and can therefore not be barcoded efficiently using the comparatively inefficient PTS protocol.

DMPS can also be used to barcode any other double-stranded template. The main disadvantage of directly barcoding and sequencing multiplex PCR products is that the optimal PCR cycle number for the multiplex PCR is strongly depending on the respective sample and can only be determined by sequencing PCR products obtained using varying cycle numbers. Therefore, the optimization of direct multiplex sequencing is less cost efficient than sequencing pools of pre-quantified singleplex PCR products.

Illumina Barcoding

All the above barcoding strategies introduce barcoding sequences at the cost of target read length by adding barcodes 3' of the sequencing primer. Illumina uses a different strategy. Truncated universal sequencing adapters labeled P5 and P7 and including the sequencing primer priming site are ligated to DNA molecules. These adapters are then brought to full length by PCR amplification with tailed primers containing the sequencing primer priming site and a "crafting sequence" tail. The crafting sequence acts as the priming site for primers used to amplify single molecules in a bridge amplification during the sequencing process. In addition to the sequencing priming site and crafting sequence the P7 primer also contains the barcode in-between the sequencing sequence and the crafting sequence (Figure 2 (d)). Barcodes are read in an independent read starting from the P7 sequencing site. As a result, the target read is not affected by the barcode. A further advantage is that non-barcoded libraries with truncated universal adapters can be stored and barcoded independently for each experiment. Thus, unlike with all other barcoding strategies presented above, libraries are not assigned a fixed barcode, allowing for greater flexibility in experimental setup.

CONCLUSION

Next-generation-sequencing of aDNA is associated with a number of problems and requires modification of existing protocols. Nevertheless, the large amounts of data produced by the various NGS instruments together with the short read length makes this new technology ideal for aDNA research. This is evidenced by the massive increase of available sequence data from long-dead organisms since the invention of NGS [58]. Due to the small amount of endogenous DNA and high background contamination characteristic for aDNA extracts, shotgun sequencing of aDNA is still only of limited use, and is mainly used for complete genome sequencing projects. Therefore, aDNA is likely to especially profit from new developments in barcoded and targeted sequencing. Capture methods in particular are a very promising approach. As the technology improves it is likely that even whole genome capture will be possible in the not too distant future, which will greatly reduce the costs and time for whole genome sequencing from extinct organisms. Next-generation-sequencing techniques carry the promise to bring aDNA research into the center of evolutionary biology and make it a crucial part of modern genetics. After all, what could help understand evolution better than following its progress in real time?

ACKNOWLEDGMENTS

We would like to thank Knut Finstermeier for producing the figures and Ann Horsburgh, Andrew Clarke and Gabrielle Beans-Picón for help with editing the manuscript. We would also like to thank two anonymous reviewers for valuable comments that improved the manuscript. M.K. is supported by the Allan Wilson Centre for Molecular Ecology and Evolution, M.H. by The University of York.

REFERENCES AND NOTES

1. Higuchi, R.; Bowman, B.; Freiberger, M.; Ryder, O.A.; Wilson, A.C. DNA-sequences from the quagga, an extinct member of the horse family. *Nature* **1984**, *312*, 282–284.
2. Saiki, R.K.; Scharf, S.; Faloona, F.; Mullis, K.B.; Horn, G.T.; Erlich, H.A.; Arnheim, N. Enzymatic amplification of beta-globin genomic sequences and restriction site analysis for diagnosis of sickle-cell anemia. *Science* **1985**, *230*, 1350–1354.
3. Mullis, K.B.; Faloona, F.A. Specific synthesis of DNA invitro via a polymerase-catalyzed chain-reaction. *Methods Enzymol.* **1987**, *155*, 335–350.
4. Saiki, R.K.; Gelfand, D.H.; Stoffel, S.; Scharf, S.J.; Higuchi, R.; Horn, G.T.; Mullis, K.B.; Erlich, H.A. Primer-directed enzymatic amplification of DNA with a thermostable DNA-polymerase. *Science* **1988**, *239*, 487–491.
5. Margulies, M.; Egholm, M.; Altman, W.E.; Attiya, S.; Bader, J.S.; Bemben, L.A.; Berka, J.; Braverman, M.S.; Chen, Y.J.; Chen, Z.T.; *et al.* Genome sequencing in microfabricated high-density picolitre reactors. *Nature* **2005**, *437*, 376–380.
6. Poinar, H.N.; Schwarz, C.; Qi, J.; Shapiro, B.; MacPhee, R.D.E.; Buigues, B.; Tikhonov, A.; Huson, D.H.; Tomsho, L.P.; Auch, A.; Rampp, M.; Miller, W.; Schuster, S.C. Metagenomics to paleogenomics: Large-scale sequencing of mammoth DNA. *Science* **2006**, *311*, 392–394.
7. Noonan, J.P.; Hofreiter, M.; Smith, D.; Priest, J.R.; Rohland, N.; Rabeder, G.; Krause, J.; Detter, J.C.; Pääbo, S.; Rubin, E.M. Genomic sequencing of Pleistocene cave bears. *Science* **2005**, *309*, 597–600.
8. Miller, W.; Drautz, D.I.; Ratan, A.; Pusey, B.; Qi, J.; Lesk, A.M.; Tomsho, L.P.; Packard, M.D.; Zhao, F.Q.; Sher, A.; Tikhonov, A.; Raney, B.; Patterson, N.; Lindblad-Toh, K.; Lander, E.S.; Knight, J.R.; Irzyk, G.P.; Fredrikson, K.M.; Harkins, T.T.; Sheridan, S.; Pringle, T.; Schuster, S.C. Sequencing the nuclear genome of the extinct woolly mammoth . *Nature* **2008**, *456*, 387–U351.
9. Green, R.E.; Krause, J.; Briggs, A.W.; Maricic, T.; Stenzel, U.; Kircher, M.; Patterson, N.; Li, H.; Zhai, W.W.; Fritz, M.H.Y.; *et al.*

A Draft Sequence of the Neandertal Genome. *Science* **2010**, *328*, 710–722.

10. Rasmussen, M.; Li, Y.R.; Lindgreen, S.; Pedersen, J.S.; Albrechtsen, A.; Moltke, I.; Metspalu, M.; Metspalu, E.; Kivisild, T.; Gupta, R.; *et al.* Ancient human genome sequence of an extinct Palaeo-Eskimo. *Nature* **2010**, *463*, 757–762.
11. Ng, S.B.; Turner, E.H.; Robertson, P.D.; Flygare, S.D.; Bigham, A.W.; Lee, C.; Shaffer, T.; Wong, M.; Bhattacharjee, A.; Eichler, E.E.; Bamshad, M.; Nickerson, D.A.; Shendure, J. Targeted capture and massively parallel sequencing of 12 human exomes . *Nature* **2009**, *461*, 272–U153.
12. Briggs, A.W.; Good, J.M.; Green, R.E.; Krause, J.; Maricic, T.; Stenzel, U.; Lalueza-Fox, C.; Rudan, P.; Brajković, D.; Kucan, Ž.; Gušić, I.; Schmitz, R.; Doronichev, V.B.; Golovanova, L.V.; de la Rasilla, M.; Fortea, J.; Rosas, A.; Pääbo, S. Targeted Retrieval and Analysis of Five Neandertal mtDNA Genomes . *Science* **2009**, *325*, 318–321.
13. Krause, J.; Briggs, A.W.; Kircher, M.; Maricic, T.; Zwyns, N.; Derevianko, A.; Pääbo, S. A Complete mtDNA Genome of an Early Modern Human from Kostenki, Russia. *Curr. Biol.* **2010**, *20*, 231–236.
14. Krause, J.; Fu, Q.M.; Good, J.M.; Viola, B.; Shunkov, M.V.; Derevianko, A.P.; Pääbo, S. The complete mitochondrial DNA genome of an unknown hominin from southern Siberia. *Nature* **2010**, *464*, 894–897.
15. Burbano, H.A.; Hodges, E.; Green, R.E.; Briggs, A.W.; Krause, J.; Meyer, M.; Good, J.M.; Maricic, T.; Johnson, P.L.F.; Xuan, Z.Y.; Rooks, M.; Bhattacharjee, A.; Brizuela, L.; Albert, F.W.; de la Rasilla, M.; Fortea, J.; Rosas, A.; Lachmann, M.; Hannon, G.J.; Pääbo, S. Targeted Investigation of the Neandertal Genome by Array-Based Sequence Capture. *Science* **2010**, *328*, 723–725.
16. Millar, C.D.; Huynen, L.; Subramanian, S.; Mohandesan, E.; Lambert, D.M. New developments in ancient genomics. *Trends Ecol. Evol.* **2008**, *23*, 386–393.
17. Poinar, H.; Kuch, M.; McDonald, G.; Martin, P.; Pääbo, S. Nuclear gene sequences from a late Pleistocene sloth coprolite. *Curr. Biol.*

2003, *13*, 1150–1152.

18. Lambert, D.M.; Ritchie, P.A.; Millar, C.D.; Holland, B.; Drummond, A.J.; Baroni, C. Rates of evolution in ancient DNA from Adelie penguins. *Science* **2002**, *295*, 2270–2273.
19. Shapiro, B.; Drummond, A.J.; Rambaut, A.; Wilson, M.C.; Matheus, P.E.; Sher, A.V.; Pybus, O.G.; Gilbert, M.T.P.; Barnes, I.; Binladen, J.; Willerslev, E.; Hansen, A.J.; Baryshnikov, G.F.; Burns, J.A.; Davydov, S.; Driver, J.C.; Froese, D.G.; Harington, C.R.; Keddie, G.; Kosintsev, P.; Kunz, M.L.; Martin, L.D.; Stephenson, R.O.; Storer, J.; Tedford, R.; Zimov, S.; Cooper, A. Rise and fall of the Beringian steppe bison. *Science* **2004**, *306*, 1561–1565.
20. Krause, J.; Dear, P.H.; Pollack, J.L.; Slatkin, M.; Spriggs, H.; Barnes, I.; Lister, A.M.; Ebersberger, I.; Pääbo, S.; Hofreiter, M. Multiplex amplification of the mammoth mitochondrial genome and the evolution of Elephantidae. *Nature* **2006**, *439*, 724–727.
21. Debruyne, R.; Schwarz, C.; Poinar, H. Comment on "Whole-genome shotgun sequencing of mitochondria from ancient hair shafts" . *Science* **2008**, *322*, 857b.
22. Maricic, T.; Pääbo, S. Optimization of 454 sequencing library preparation from small amounts of DNA permits sequence determination of both DNA strands. *Biotechniques* **2009**, *46*, 51–57.
23. Meyer, M.; Kircher, M. Illumina Sequencing Library Preparation for Highly Multiplexed Target Capture and Sequencing . *Cold Spring Harb. Protoc.* **2010**.
24. Green, R.E.; Krause, J.; Ptak, S.E.; Briggs, A.W.; Ronan, M.T.; Simons, J.F.; Du, L.; Egholm, M.; Rothberg, J.M.; Paunovic, M.; Pääbo, S. Analysis of one million base pairs of Neanderthal DNA. *Nature* **2006**, *444*, 330–336.
25. Gilbert, M.T.P.; Tomsho, L.P.; Rendulic, S.; Packard, M.; Drautz, D.I.; Sher, A.; Tikhonov, A.; Dalén, L.; Kuznetsova, T.; Kosintsev, P.; Campos, P.F.; Higham, T.; Collins, M.J.; Wilson, A.S.; Shidlovskiy, F.; Buigues, B.; Ericson, P.G.P.; Germonpré, M.; Götherström, A.; Iacumin, P.; Nikolaev, V.; Nowak-Kemp, M.; Willerslev, E.; Knight, J.R.; Irzyk, G.P.; Perbost, C.S.; Fredrikson, K.M.; Harkins, T.T.; Sheridan, S.; Miller, W.; Schuster, S.C. Whole-genome shotgun sequencing of mitochondria from ancient hair shafts. *Science* **2007**,

317, 1927–1930.

26. Gilbert, M.T.P.; Drautz, D.I.; Lesk, A.M.; Ho, S.Y.W.; Qi, J.; Ratan, A.; Hsu, C.H.; Sher, A.; Dalén, L.; Götherström, A.; *et al.* Intraspecific phylogenetic analysis of Siberian woolly mammoths using complete mitochondrial genomes. *Proc. Natl. Acad. Sci. U. S. A.* **2008**, *105*, 8327–8332.
27. Gilbert, M.T.P.; Kivisild, T.; Grønnow, B.; Andersen, P.K.; Metspalu, E.; Reidla, M.; Tamm, E.; Axelsson, E.; Götherström, A.; Campos, P.F.; Rasmussen, M.; Metspalu, M.; Higham, T.F.G.; Schwenninger, J.L.; Nathan, R.; De Hoog, C.J.; Koch, A.; Møller, L.N.; Andreasen, C.; Meldgaard, M.; Villems, R.; Bendixen, C.; Willerslev, E. Paleo-Eskimo mtDNA genome reveals matrilineal discontinuity in Greenland. *Science* **2008**, *320*, 1787–1789.
28. Green, R.E.; Malaspinas, A.S.; Krause, J.; Briggs, A.W.; Johnson, P.L.F.; Uhler, C.; Meyer, M.; Good, J.M.; Maricic, T.; Stenzel, U.; Prüfer, K.; Siebauer, M.; Burbano, H.A.; Ronan, M.; Rothberg, J.M.; Egholm, M.; Rudan, P.; Brajković, D.; Kućan, Ž.; Gušić, I.; Wikström, M.; Laakkonen, L.; Kelso, J.; Slatkin, M.; Pääbo, S. A complete neandertal mitochondrial genome sequence determined by high-throughput Sequencing . *Cell* **2008**, *134*, 416–426.
29. Miller, W.; Drautz, D.I.; Janecka, J.E.; Lesk, A.M.; Ratan, A.; Tomsho, L.P.; Packard, M.; Zhang, Y.; McClellan, L.R.; Qi, J.; Zhao, F.; Gilbert, M.T.P.; Dalén, L.; Arsuaga, J.L.; Ericson, P.G.P.; Huson, D.H.; Helgen, K.M.; Murphy, W.J.; Götherström, A.; Schuster, S.C. The mitochondrial genome sequence of the Tasmanian tiger (Thylacinus cynocephalus). *Genome Res.* **2009**, *19*, 213–220.
30. Allentoft, M.E.; Schuster, S.C.; Holdaway, R.N.; Hale, M.L.; McLay, E.; Oskam, C.; Gilbert, M.T.P.; Spencer, P.; Willerslev, E.; Bunce, M. Identification of microsatellites from an extinct moa species using high-throughput (454) sequence data. *Biotechniques* **2009**, *46*, 195–200.
31. Ramírez, O.; Gigli, E.; Bover, P.; Alcover, J.A.; Bertranpetit, J.; Castresana, J.; Lalueza-Fox, C. Paleogenomics in a Temperate Environment: Shotgun Sequencing from an Extinct Mediterranean Caprine . *Plos One* **2009**, *4*, e5670.
32. Willerslev, E.; Gilbert, M.T.P.; Binladen, J.; Ho, S.Y.W.; Campos,

P.F.; Ratan, A.; Tomsho, L.P.; da Fonseca, R.R.; Sher, A.; Kuznetsova, T.V.; Nowak-Kemp, M.; Roth, T.L.; Miller, W.; Schuster, S.C. Analysis of complete mitochondrial genomes from extinct and extant rhinoceroses reveals lack of phylogenetic resolution . *BMC Evol. Biol.* **2009**, *9*, 95.

33. Zhao, F.Q.; Qi, J.; Schuster, S.C. Tracking the past: Interspersed repeats in an extinct Afrotherian mammal, Mammuthus primigenius. *Genome Res.* **2009**, *19*, 1384–1392.
34. Stiller, M.; Knapp, M.; Stenzel, U.; Hofreiter, M.; Meyer, M. Direct multiplex sequencing (DMPS)-a novel method for targeted high-throughput sequencing of ancient and highly degraded DNA. *Genome Res.* **2009**, *19*, 1843–1848.
35. Edwards, C.J.; Magee, D.A.; Park, S.D.E.; McGettigan, P.A.; Lohan, A.J.; Murphy, A.; Finlay, E.K.; Shapiro, B.; Chamberlain, A.T.; Richards, M.B.; Bradley, D.G.; Loftus, B.J.; MacHugh, D.E. A Complete Mitochondrial Genome Sequence from a Mesolithic Wild Aurochs (Bos primigenius) . *Plos One* **2010**, *5*, e9255.
36. Lindqvist, C.; Schuster, S.C.; Sun, Y.Z.; Talbot, S.L.; Qi, J.; Ratan, A.; Tomsho, L.P.; Kasson, L.; Zeyl, E.; Aars, J.; Miller, W.; Ingólfsson, Ó.; Bachmann, L.; Wiig, Ø. Complete mitochondrial genome of a Pleistocene jawbone unveils the origin of polar bear . *Proc. Natl. Acad. Sci. U. S. A.* **2010**, *107*, 5053–5057.
37. Ho, S.Y.W.; Gilbert, M.T.P. Ancient mitogenomics. *Mitochondrion (Kidlington)* **2010**, *10*, 1–11.
38. Sanger, F.; Nicklen, S.; Coulson, A.R. DNA sequencing with chain-terminating inhibitors. *Proc. Natl. Acad. Sci. U. S. A.***1977**, *74*, 5463–5467.
39. Meyer, M.; Stenzel, U.; Myles, S.; Prüfer, K.; Hofreiter, M. Targeted high-throughput sequencing of tagged nucleic acid samples . *Nucleic Acids Res.* **2007**, *35*, No. 15 e97.
40. Römpler, H.; Dear, P.H.; Krause, J.; Meyer, M.; Rohland, N.; Schöneberg, T.; Spriggs, H.; Stiller, M.; Hofreiter, M. Multiplex amplification of ancient DNA. *Nat. Protoc.* **2006**, *1*, 720–728.
41. Römpler, H.; Rohland, N.; Lalueza-Fox, C.; Willerslev, E.; Kuznetsova, T.; Rabeder, G.; Bertranpetit, J.; Schöneberg, T.; Hofreiter, M. Nuclear gene indicates coat-color polymorphism in

mammoths. *Science* **2006**, *313*, 62–62.

42. Rohland, N.; Malaspinas, A.S.; Pollack, J.L.; Slatkin, M.; Matheus, P.; Hofreiter, M. Proboscidean mitogenomics: Chronology and mode of elephant evolution using mastodon as outgroup. *PLoS Biol.* **2007**, *5*, 1663–1671.
43. Krause, J.; Unger, T.; Noçon, A.; Malaspinas, A.S.; Kolokotronis, S.O.; Stiller, M.; Soibelzon, L.; Spriggs, H.; Dear, P.H.; Briggs, A.W.; Bray, S.C.E.; O'Brien, S.J.; Rabeder, G.; Matheus, P.; Cooper, A.; Slatkin, M.; Pääbo, S.; Hofreiter, M. Mitochondrial genomes reveal an explosive radiation of extinct and extant bears near the Miocene-Pliocene boundary . *BMC Evol. Biol.* **2008**, *8*, 220.
44. Prüfer, K.; Stenzel, U.; Hofreiter, M.; Pääbo, S.; Kelso, J.; Green, R.E. Computational challenges in the analysis of ancient DNA . *Genome Biology* **2010**, *11*, R47.
45. Hofreiter, M.; Jaenicke, V.; Serre, D.; von Haeseler, A.; Pääbo, S. DNA sequences from multiple amplifications reveal artifacts induced by cytosine deamination in ancient DNA. *Nucleic Acids Res.* **2001**, *29*, 4793–4799.
46. Willerslev, E.; Hansen, A.J.; Binladen, J.; Brand, T.B.; Gilbert, M.T.P.; Shapiro, B.; Bunce, M.; Wiuf, C.; Gilichinsky, D.A.; Cooper, A. Diverse plant and animal genetic records from Holocene and Pleistocene sediments. *Science* **2003**, *300*, 791–795.
47. Hofreiter, M.; Mead, J.I.; Martin, P.; Poinar, H.N. Molecular caving . *Curr. Biol.* **2003**, *13*, R693–R695.
48. Willerslev, E.; Cappellini, E.; Boomsma, W.; Nielsen, R.; Hebsgaard, M.B.; Brand, T.B.; Hofreiter, M.; Bunce, M.; Poinar, H.N.; Dahl-Jensen, D.; Johnsen, S.; Steffensen, J.P.; Bennike, O.; Schwenninger, J.L.; Nathan, R.; Armitage, S.; de Hoog, C.J.; Alfimov, V.; Christl, M.; Beer, J.; Muscheler, R.; Barker, J.; Sharp, M.; Penkman, K.E.H.; Haile, J.; Taberlet, P.; Gilbert, M.T.P.; Casoli, A.; Campani, E.; Collins, M.J. Ancient biomolecules from deep ice cores reveal a forested Southern Greenland. *Science* **2007**, *317*, 111–114.
49. Albert, T.J.; Molla, M.N.; Muzny, D.M.; Nazareth, L.; Wheeler, D.; Song, X.Z.; Richmond, T.A.; Middle, C.M.; Rodesch, M.J.; Packard, C.J.; Weinstock, G.M.; Gibbs, R.A. Direct selection of human genomic loci by microarray hybridization. *Nat. Meth.* **2007**,

4, 903–905.

50. Hodges, E.; Xuan, Z.; Balija, V.; Kramer, M.; Molla, M.N.; Smith, S.W.; Middle, C.M.; Rodesch, M.J.; Albert, T.J.; Hannon, G.J.; McCombie, W.R. Genome-wide in situ exon capture for selective resequencing. *Nat. Genet.* **2007**, *39*, 1522–1527.
51. Gnirke, A.; Melnikov, A.; Maguire, J.; Rogov, P.; LeProust, E.M.; Brockman, W.; Fennell, T.; Giannoukos, G.; Fisher, S.; Russ, C.; Gabriel, S.; Jaffe, D.B.; Lander, E.S.; Nusbaum, C. Solution hybrid selection with ultra-long oligonucleotides for massively parallel targeted sequencing. *Nat. Biotechnol.* **2009**, *27*, 182–189.
52. Hodges, E.; Rooks, M.; Xuan, Z.Y.; Bhattacharjee, A.; Gordon, D.B.; Brizuela, L.; McCombie, W.R.; Hannon, G.J. Hybrid selection of discrete genomic intervals on custom-designed microarrays for massively parallel sequencing. *Nat. Protoc.* **2009**, *4*, 960–974.
53. Mamanova, L.; Coffey, A.J.; Scott, C.E.; Kozarewa, I.; Turner, E.H.; Kumar, A.; Howard, E.; Shendure, J.; Turner, D.J. Target-enrichment strategies for next-generation sequencing. *Nat. Meth.* **2010**, *7*, 111–118.
54. Noonan, J.P.; Coop, G.; Kudaravalli, S.; Smith, D.; Krause, J.; Alessi, J.; Platt, D.; Pääbo, S.; Pritchard, J.K.; Rubin, E.M. Sequencing and analysis of Neanderthal genomic DNA. *Science* **2006**, *314*, 1113–1118.
55. Anderung, C.; Persson, P.; Bouwman, A.; Elburg, R.; Götherström, A. Fishing for ancient DNA. *Forensic Sci. Int.: Genetics* **2008**, *2*, 104–107.
56. Briggs, A.W.; Good, J.M.; Green, R.E.; Krause, J.; Maricic, T.; Stenzel, U.; Pääbo, S. Primer Extension Capture: Targeted Sequence Retrieval from Heavily Degraded DNA Sources . *J. Vis. Exp.* **2009**, *31*.
57. Meyer, M.; Stenzel, U.; Hofreiter, M. Parallel tagged sequencing on the 454 platform. *Nat. Protoc.* **2008**, *3*, 267–278.
58. Shapiro, B.; Hofreiter, M. Analysis of ancient human genomes. *Bioessays* **2010**, *32*, 388–391.

CITATIONS

CHAPTER 2

Lin Liu, Yinhu Li, Siliang Li, et al., "Comparison of Next-Generation Sequencing Systems," Journal of Biomedicine and Biotechnology, vol. 2012, Article ID 251364, 11 pages, 2012. doi:10.1155/2012/251364.

CHAPTER 3

Ansorge WJ (2016) Next Generation DNA Sequencing (II): Techniques, Applications. Next Generat Sequenc & Applic S1:005. doi:10.4172/2469-9853.S1-005.

CHAPTER 4

Attia TH, Saeed MA (2016) Next Generation Sequencing Technologies: A Short Review. Next Generat Sequenc & Applic S1:006. doi:10.4172/2469-9853.S1-006.

CHAPTER 5

Pareek CS (2015) Transcriptome Analysis on RNA-seq Data. Next Generat Sequenc & Applic S1:002. doi:10.4172/2469-9853.S1-002.

CHAPTER 6

Li WL (2016) Interpretation of Clinical Next-Generation Sequencing Data: A Hurdle to Jump Over. Next Generat Sequenc & Applic 3:130. doi:10.4172/2469-9853.1000130.

CHAPTER 7

Suma J, Sivakumar G (2016) Next Generation Sequencing-Current Status. Next Generat Sequenc & Applic 3:e107. doi: 0.4172/2469-9853.3-e107.

CHAPTER 8

Nawab DH (2015) The Pharmaceutical Applications of Next Generation Sequencing in Oncology Drug Designing and Development. Next Generat Sequenc & Applic 2:116. doi:10.4172/2469-9853.1000116.

CHAPTER 9

Blum HE, Oexle K (2014) Clinical Interpretation and Implications of Whole Genome Sequencing. Next Generat Sequenc & Applic 1:105. doi: 10.4172/2469-9853.1000105.

CHAPTER 10

Fox EJ, Reid-Bayliss KS, Emond MJ, Loeb LA (2014) Accuracy of Next Generation Sequencing Platforms. Next Generat Sequenc & Applic 1: 106. doi: 1000106 10.4172/2469-9853.1000106.

CHAPTER 11

Srinivasan S, Batra J (2014) Four Generations of Sequencing- Is it Ready for the Clinic Yet?. Next Generat Sequenc & Applic 1:107. doi:10.4172/2469-9853.1000107.

CHAPTER 12

Govindan VP, Murthy KP (2016) Next Generation Sequencing Technologies in Malaria. Next Generat Sequenc & Applic 3:138. doi: 10.4172/2469-9853.1000138.

CHAPTER 13

Munoz-Minjares J, Shmaliy YS (2016) The Role of Optimal Detection of CNAs and Error Analysis Using Next Generation Sequencing. Next Generat Sequenc & Applic 3: 141. doi:10.4172/2469-9853.1000141.

CHAPTER 14

Ku CS (2016) Bringing Next-Generation Sequencing Oncology Tests into the Diagnostic Setting. Next Generat Sequenc & Applic S1:004. doi:10.4172/2469-9853.S1-004.

CHAPTER 15

Blue GM, Winlaw DS (2015) Next Generation Sequencing in Congenital Heart Disease: Gene Discovery and Clinical Application. Next Generat Sequenc & Applic 2:113. doi:10.4172/2469-9853.1000113.

CHAPTER 16

Mabey T, Honsawek S (2014) Genome Wide Association Studies and Next Generation Sequencing Technologies in Osteoarthritis. Next Generat Sequenc & Applic 1:108. doi:10.4172/2469-9853.1000108.

CHAPTER 17

Ye F, David C, Samuels, Guo Y (2014) Detection of Mitochondrial Mutations in Cancer by Next Generation Sequencing. Next Generat Sequenc & Applic 1:102. doi:10.4172/2469-9853.1000102.

CHAPTER 18

Kristen H. Taylor, Huidong Shi and Charles W. Caldwell, "Next Generation Sequencing: Advances in Characterizing the Methylome," doi:10.3390/genes1020143.

CHAPTER 19

Korneliussen et al.: ANGSD: Analysis of Next Generation Sequencing Data. BMC Bioinformatics 2014 15:356. doi:10.1186/s12859-014-0356-4.

CHAPTER 20

Michael Knapp and Michael Hofreiter, "Next Generation Sequencing of Ancient DNA: Requirements, Strategies and Perspectives," doi:10.3390/genes1020227.

INDEX